State-of-the-Art and Emerging Technologies for Therapeutic Monoclonal Antibody Characterization
Volume 1. Monoclonal Antibody Therapeutics: Structure, Function, and Regulatory Space

ACS SYMPOSIUM SERIES 1176

State-of-the-Art and Emerging Technologies for Therapeutic Monoclonal Antibody Characterization
Volume 1. Monoclonal Antibody Therapeutics: Structure, Function, and Regulatory Space

John E. Schiel, Editor
National Institute of Standards and Technology
Gaithersburg, Maryland

Darryl L. Davis, Editor
Janssen Research and Development, LLC
Spring House, Pennsylvania

Oleg V. Borisov, Editor
Novavax, Inc.
Gaithersburg, Maryland

American Chemical Society, Washington, DC
Distributed in print by Oxford University Press

Library of Congress Cataloging-in-Publication Data

State-of-the-art and emerging technologies for therapeutic monoclonal antibody characterization / John E. Schiel, editor, National Institute of Standards and Technology, Gaithersburg, Maryland, Darryl L. Davis, editor, Janssen Research and Development, LLC, Spring House, Pennsylvania, Oleg V. Borisov, editor, Novavax, Inc., Gaithersburg, Maryland.
    volumes cm. -- (ACS symposium series ; 1176)
  Includes bibliographical references and index.
  Contents: v. 1. monoclonal antibody therapeutics : structure, function, and regulatory space
  ISBN 978-0-8412-3026-2
  1. Monoclonal antibodies. 2. Immunoglobulins--Therapeutic use. I. Schiel, John E., editor. II. Davis, Darryl L., editor. III. Borisov, Oleg V., editor.
  QR186.85.S73 2014
  616.07'98--dc23
                             2014040141

The paper used in this publication meets the minimum requirements of American National Standard for Information Sciences—Permanence of Paper for Printed Library Materials, ANSI Z39.48n1984.

Copyright © 2014 American Chemical Society

Distributed in print by Oxford University Press

All Rights Reserved. Reprographic copying beyond that permitted by Sections 107 or 108 of the U.S. Copyright Act is allowed for internal use only, provided that a per-chapter fee of $40.25 plus $0.75 per page is paid to the Copyright Clearance Center, Inc., 222 Rosewood Drive, Danvers, MA 01923, USA. Republication or reproduction for sale of pages in this book is permitted only under license from ACS. Direct these and other permission requests to ACS Copyright Office, Publications Division, 1155 16th Street, N.W., Washington, DC 20036.

The citation of trade names and/or names of manufacturers in this publication is not to be construed as an endorsement or as approval by ACS of the commercial products or services referenced herein; nor should the mere reference herein to any drawing, specification, chemical process, or other data be regarded as a license or as a conveyance of any right or permission to the holder, reader, or any other person or corporation, to manufacture, reproduce, use, or sell any patented invention or copyrighted work that may in any way be related thereto. Registered names, trademarks, etc., used in this publication, even without specific indication thereof, are not to be considered unprotected by law.

PRINTED IN THE UNITED STATES OF AMERICA

Foreword

The ACS Symposium Series was first published in 1974 to provide a mechanism for publishing symposia quickly in book form. The purpose of the series is to publish timely, comprehensive books developed from the ACS sponsored symposia based on current scientific research. Occasionally, books are developed from symposia sponsored by other organizations when the topic is of keen interest to the chemistry audience.

Before agreeing to publish a book, the proposed table of contents is reviewed for appropriate and comprehensive coverage and for interest to the audience. Some papers may be excluded to better focus the book; others may be added to provide comprehensiveness. When appropriate, overview or introductory chapters are added. Drafts of chapters are peer-reviewed prior to final acceptance or rejection, and manuscripts are prepared in camera-ready format.

As a rule, only original research papers and original review papers are included in the volumes. Verbatim reproductions of previous published papers are not accepted.

ACS Books Department

Contents

Preface ix

1. **Monoclonal Antibody Therapeutics: The Need for Biopharmaceutical Reference Materials** 1
   John E. Schiel, Anthony Mire-Sluis, and Darryl Davis

2. **Monoclonal Antibodies: Mechanisms of Action** 35
   Roy Jefferis

3. **Heterogeneity of IgGs: Role of Production, Processing, and Storage on Structure and Function** 69
   Chris Barton, David Spencer, Sophia Levitskaya, Jinhua Feng, Reed Harris, and Mark A. Schenerman

4. **Perspectives on Well-Characterized Biological Proteins** 99
   Kurt Brorson and Brent Kendrick

5. **Using Quality by Design Principles in Setting a Control Strategy for Product Quality Attributes** 117
   Gregory C. Flynn and Gregg B. Nyberg

Appendix 151

Editors' Biographies 157

Indexes

Author Index 161

Subject Index 163

Preface

Monoclonal antibodies (mAbs) are an important class of therapeutic that has greatly expanded our ability to treat a variety of indications, including cancer, autoimmune disorders, and infectious diseases. The cost of developing these molecules and ensuring that they are fit for purpose is much higher than for small-molecule drugs. Although some of these costs are due in part to the nature of production, a major cost is incurred because of the complexity of the molecule itself. Extensive analytical testing regimes are needed for in-depth characterization and to ensure product stability, proper in-process controls, safety, and efficacy. The next generation of product understanding will require highly complex, orthogonal technologies to elucidate interdependent structure-function relationships.

Despite high development costs, the growth of mAb therapeutics continues to soar, with sales of mAb-based therapeutics accounting for ~$50 billion U.S. annually and for seven of the top-ten selling biologic drugs. In anticipation of patent expiry for seven major mAb therapeutics prior to 2020, the U.S. Food and Drug Administration (FDA) released draft guidance on biosimilar approval in 2012. A landmark approval of the first mAb biosimilar occurred in 2013 when the European Commission approved two biosimilar versions of infliximab. It appears likely that the extensive pipeline of biosimilar products will soon become a commercial reality on a global scale. The need for critical assessment of therapeutic mAb characterization is therefore at a precipice, as it is expected that analytical and biophysical characterization strategies will play an ever-increasing role in biosimilarity assessment.

Regardless of the pathway to market, the pursuit of improved public health drives therapeutic development. Ultimately, assurance of public safety is of the highest priority. Considering the entirety of a drug's lifecycle, there is no greater need than for a suite of technologies capable of verifying mechanism of actions (MoAs), identity, and the product and process consistency of such life-saving medicines. Although characterization methods for mAbs continue to mature, a holistic approach combining a multitude of scientific backgrounds is required for the characterization of various attributes of each individual drug candidate. To that end, a manufacturer-specific repository of each drug must be retained as a comparator to ensure consistent production over time. Although this approach is indispensable, it also limits the ability for cross-agency harmonization of industry best practices and limits collaboration across industry, government agencies, academia, and instrument manufacturers. The unique challenge of protein therapeutics resulted in a consolidated effort between industry, academia, and federal agencies to identify a means to better progress from drug target to therapy. A group of forward-thinking individuals testified before the U.S. House of Representatives Committee on Science and Technology in 2009 on behalf of the biotherapeutic community regarding the need for better tools and the role that standards can play. The seed was planted for a concerted effort toward the development of appropriate standards and has continued to grow through numerous workshops, seminars, and round-table discussions. Throughout this time, it has become clear that researchers and regulators share a passion for producing the highest quality products through controlled production and robust characterization.

The rigors of biotherapeutic development and analysis have clearly indicated a need for control over every stage of development. Critical evaluation of process steps and final product requires technique-specific standards to supplement the in-house repository of specific drug substances. Constant dialogue between industry, regulatory bodies, and standards organizations has identified the need for standards and associated data to better define method performance. The informal consortium referenced above, along with years of testing potential method-specific standards, has shown that a class-specific molecule that embodies all of the technical challenges of that class is necessary. Therefore, an IgG1κ was selected as the premier target capable of covering the broadest range of applications and drug development targets. The mere presence of such a standard will not guarantee success; it will take the unified effort of all involved to incorporate such a material into an already robust pipeline. The NISTmAb IgG1κ, as described throughout this series, is being introduced as a collaborative tool to critically evaluate current characterization strategies. The proposed reference material shows promise as a mechanism for accelerating next-generation technology into the routine development environment.

Scientific and health care advances have arisen throughout history, due in part to both directed scientific approaches and a small dose of luck. Inevitably, it is the collaborative effort of many who have advanced this field to a refined state capable of meeting the global capacity and needs of the populous. The current book series represents a defining moment in the development of technologies for characterizing mAbs. A multitude of experts in the field have come together around a single molecule for the first time to demonstrate both current and future practices involved in the characterization of a biotherapeutic. The current series presents an open discussion of current best practices, a multitude of intuitive research, and a collaborative philosophy on where the field needs to move to satisfy future scientific and regulatory needs. Throughout this effort, a first-of-a-kind repository of regulatory considerations, experimental methods, and data—as well as a widely available "reference" mAb—are now available to industry, academia, regulatory personnel, and instrument manufacturers. We hope that this compilation serves as a baseline for many years of future collaboration, continued development, and ultimately a routine analytical pipeline for rapid time-to-market for mAb therapeutics.

The editors acknowledge Jane Ladner for preparation of the NISTmAb model structure featured on the cover. This model was built as described in the Higher Order Structure chapter/Volume 3, Chapter 2.

John E. Schiel
Research Chemist
Biomolecular Measurement Division
National Institute of Standards and Technology
Gaithersburg, Maryland 20899, United States
john.schiel@nist.gov (e-mail)

Darryl L. Davis
Associate Scientific Director
Janssen Research and Development, LLC
Spring House, Pennsyvania 19002, United States
DDavis14@its.jnj.com (e-mail)

Oleg V. Borisov
Associate Director
Novavax, Inc.
Gaithersburg, Maryland 20878, United States
oborisov@novavax.com (e-mail)

Chapter 1

Monoclonal Antibody Therapeutics: The Need for Biopharmaceutical Reference Materials

John E. Schiel,*,[1] Anthony Mire-Sluis,[2] and Darryl Davis[3]

[1]National Institute of Standards and Technology,
Biomolecular Measurement Division,
Gaithersburg, Maryland 20899, United States
[2]North America, Singapore, Abingdon, Contract and Product Quality,
Amgen Inc., Thousand Oaks, California 91320, United States
[3]Janssen Research and Development, LLC,
Spring House, Pennsylvania 19002, United States
*E-mail: john.schiel@nist.gov

Therapeutic monoclonal antibodies (mAbs) harness the highly evolved specificity of adaptive immunity to fight disease. mAb-based therapeutics have grown exponentially with the advent of mammalian cell culture, process, and formulation technology. At the same time, state-of-the-art and emerging analytical and biophysical methodology provides very detailed process and product information. Although such a battery of methodology and wealth of information is critical to product understanding, the accuracy, precision, robustness, and suitability of such techniques are also of critical importance. Performance specifications have previously been set on a product-specific basis and continued suitability verified with trending and comparability to in-house product-specific reference standards. This mechanism is likely irreplaceable due to the highly individual yet heterogeneous nature of mAb therapeutics. However, a representative and widely available material, coupled with detailed historical data, would greatly supplement characterization efforts throughout the drug product lifecycle. To this end, a first-of-its kind

© 2014 American Chemical Society

qualitative and quantitative biopharmaceutical reference material to supplement drug substance/product characterization is described. The NISTmAb IgG1κ is intended to provide a well-characterized, longitudinally available test material that is expected to greatly facilitate development of originator and follow-on biologics for the foreseeable future.

Introduction

Significant advances in modern medicine are often directly intertwined with production of novel disease treatments. The documented use of herbal remedies for ailments dates back to 3,000 B.C., when ancient Egyptian and Chinese cultures used various plants for their healing properties (*1*). Therapeutic effects of herbal medicine are a result of bioactive chemical substances, many of which have been identified and synthetically manufactured as small molecule drugs. For example, acetylsalicylic acid (Aspirin®) is a synthetic derivative of willow tree extract identified to have fewer side effects and eventually became the first blockbuster drug (*2*). Since that time, small molecule drugs have been developed for many indications and will likely continue to play a significant role in healthcare.

In addition to serving as human and animal medicines, many naturally derived drug products confer a selective advantage to the host species. In the same manner as humans have utilized this natural selection from plants, fungi, and other natural materials to derive small molecule drugs, it is a logical step to harness animal-derived immune defenses to produce therapeutics. The human immune system is comprised of both innate and adaptive immunity (*3, 4*). Innate immunity confers a rapid initial line of defense via recognition of evolutionarily conserved features from pathogenic bacteria, viruses, and other invading organisms. Innate immune responses include the epithelial layer, which serves to block pathogen entry; phagocytic cells such as neutrophils and macrophages that directly ingest and kill pathogens; and inflammatory responses (e.g., cytokines, chemokines) that assist in recruiting additional innate and/or adaptive immune responses (*3, 4*).

The adaptive immune response centers around the ability of T-cells and B-cells to form a learned response against a specific target pathogen following initial sensitization (*4–6*). Although adaptive immunity is relatively slow (days or more) to respond to initial infection, it is unique in that it remembers specific pathogenic antigens and is able to mount a more rapid and specific protection against subsequent exposure. T-cells are adaptive immune cells that recognize peptide antigens. When an organism is initially infected, phagocytic cells or infected cells will process pathogen proteins into their constituent peptides through lysosomal degradation. Pathogen peptides can then be associated with a major histocompatibility complex (MHC) and presented to the extracellular surface. These MHC-associated peptides on the antigen-presenting cell surface are recognized by T-cells, which induce apoptosis of the infected cell and/or a chemotactic response that recruits additional adaptive and innate immune functions to aid in clearance of the infection (*5*).

The second major adaptive immune response (humoral response) is reliant on B-cells expressing immunoglobulin (Ig or antibody) proteins (*4–6*). Antibodies specifically recognize regions of pathogens such as proteins, carbohydrates, or lipids that may be present on the invading organism (*4*). Immunoglobulins are divided into classes (isotypes) and subclasses based on their structure as described in more detail in the Mechanisms of Action chapter/Volume 1, Chapter 2. The different human isotypes (IgA, IgG, IgM, IgD, and IgE) each have a unique distribution and function in the adaptive immune response (*4*). All currently approved monoclonal antibody (mAb) therapeutics harness the immunological capability of the IgG-class antibody, which also happens to be the highest concentration Ig class in blood (*4*).

The naturally occurring humoral response begins with activation of a naive B-cell expressing an IgM antibody on its cell surface. Each individual B-cell produces an IgM on its cell surface that specifically targets a single antigenic site or epitope. When a circulating B-cell recognizes its particular antigen, the cell will proliferate memory and effector B-cells. Memory B-cells continue to express antigen-specific IgM, thereby conferring a long-lasting learned memory of the initial infection. Effector B-cells, on the other hand, undergo class switching and are induced to produce soluble IgG targeting the same epitope. Soluble IgG binds to circulating pathogen and leads to removal of the invading pathogen through effector-mediated functions such as complement-dependent cytotoxicity (CDC), antigen-dependent cellular cytotoxicity (ADCC), or direct clearance through Fc binding in appropriate organs (*4*). Antibodies and the humoral defense are very effective at fighting a wide range of diseases. This response mechanism can also be considered somewhat more simplistic because the antibody recognizes intact pathogen as opposed to a T-cell response via antigen-presenting cell (APC)-processed antigen. It is therefore no surprise that IgG proteins were targeted for their potential utility as therapeutics.

The first demonstration of IgG-related therapeutic efficacy dates back to 1890, when serum from rabbits immunized with tetanus toxin conferred immunity to naive animals (*7*). The first clinical use of whole human serum was in 1907 for the prevention of measles, and this treatment proved to be of great importance during the early 20th century (*8*). The Ig component of serum was quickly recognized for its role in adaptive immunity, and technology was developed to purify the Ig fraction for selective use as a therapeutic (*9, 10*). Intramuscular injection of serum Ig was initially used; however, intravenous (IV) administration was soon recognized to result in fewer infections. The use of IV Ig therapies is now approved for a variety of indications, including primary humoral immunodeficiency, B-cell chronic lymphocytic leukemia, Kawasaki disease, and bone marrow transplantation (*11*).

The therapeutic benefits derived from IV Ig, as well as the typical humoral response in animals, are polyclonal in nature. In other words, an invading organism elicits a response from numerous B-cells, and IgGs of different epitopic specificity are produced. In 1975, Kohler and Milstein first described the *in vitro* production of mAbs with specificity for a single epitope using murine hybridoma technology (*12*) and were later awarded the Nobel prize. Production of a mAb with this technique involves first sensitizing a mouse with a human antigen. Murine B-cells are then extracted from the spleen and fused with immortalized myeloma cells (a cancerous plasma cell) to form a mAb-producing hybridoma. Tissue cultures or living mice can then be used to increase production of the mAb.

Due to the highly selective nature of a given mAb, mAbs of a given primary amino acid sequence can be thought of as unique entities. Therapeutic mAbs are, therefore, individually named, typically with both a trademarked name (trade name) as well as a nonproprietary name based on the accepted International Nonproprietary Names (INN) Programme (13, 14). INN nomenclature consists of a sufficiently distinctive prefix, a series of substems, and a suffix in the form of "Prefix-SubstemA-SubstemB-suffix." The suffix "-mab" is common to all nonproprietary names. Substem A and substem B indicate the antigen target class and the species on which the immunoglobulin sequence is based, respectively, as described in Table 1.

Table 1. System for International Nonproprietary Naming of Monoclonal Antibody (mAb) Therapeutics*

Prefix	Substem A		Substem B		Suffix
The prefix must be a unique, distinctive name	-b(a)-	Bacterial	a	Rat	-mAb
	-c(i)-	Cardiovascular	axo	Rat-mouse	
	-f(u)-	Fungal	e	Hamster	
	-k(i)-	Interleukin	i	Primate	
	-l(i)-	Immunomodulating	o	Mouse	
	-n(e)-	Neural	u	Human	
	-s(o)-	Bone	xi	Chimeric	
	-tox(a)-	Toxin	xizu	Chimeric-humanized	
	-t(u)-	Tumor			
	-v(i)-	Viral	zu	Humanized	

* Substem A represents the classification of the mAbs antigenic specificity, and substem B represents the species upon which the primary amino acid sequence is based.

The first murine (-omab) hybridoma-produced mAb therapeutic was realized in 1986 with the U.S. Food and Drug Administration (FDA) market approval of Orthoclone® (muronomab) (15, 16). Interest in mAb therapies rapidly grew due to their potential for a long half-life (as a result of catabolic recycling described in the Mechanisms of Action chapter/Volume 1, Chapter 2) and their unsurpassed specificity. However, extraction of therapeutic mAbs from mouse ascites fluid via hybridoma technology did not yield a large number of approved therapeutics due to the need for animal hosts as well as insufficient titers to support drug development (17). Their murine origin was also quickly identified to result in non-self recognition of idiotypic determinants by the human immune system as well as a less than optimal elucidation of effector functions (18, 19).

Recombinant DNA technology resolved many difficulties associated with the production of protein therapeutics using animal hosts for therapeutic expression. Production of protein therapeutics via recombinant DNA technology begins with a cloning vector (e.g., plasmid or viral DNA). The desired sequence encoding the protein therapeutic, a promoter, and a selection marker sequence is ligated with the vector to form appropriate recombinant DNA. Recombinant DNA can then be transfected, or inserted, into the host cell DNA of a suitable expression system containing the molecular machinery required for replication (*20–22*). Successfully transfected host cells are selected through growth in a medium requiring expression of metabolic-selectable markers or antibiotic-selectable markers for cell viability (*20*). Further clonal selection can also be undertaken to obtain a population optimized for characteristics such as cell line stability, product yield, and product quality (*20*). Through the years, there have been a number of advances in gene integration, as well as clonal selection, which have been recently reviewed (*22, 23*). Selected cells contain incorporated DNA that encodes the product, as well as a promoter sequence capable of inducing high levels of transcription and, therefore, protein therapeutic production. The ability to insert "your favorite gene" also paved the way for introduction of sequences encoding for more human-like DNA.

Chimeric antibodies were the first recombinant therapeutics developed in an effort to reduce immunogenic responses and improve effector functions compared to fully murine mAbs (*24, 25*). Chimeric antibodies (-ximab), first demonstrated in 1984, consist of a human constant region spliced with a fully murine variable region (*24*). The "self" Fc domain resulted in longer half-life and a higher propensity to elicit the Fc effector functions that are critical to certain modes of action, as described in the Mechanisms of Action chapter/Volume 1, Chapter 2). The first approved chimeric product was Reopro®, a chimeric monoclonal antibody antigen-binding fragment (Fab) for the prevention of ischemic complications during angioplasty (*25–27*). Despite potential for non-self immunogenic responses to the remaining murine component, numerous chimeric intact antibodies have also been approved, including Rituxan® and Erbitux® (anticancer agents), and Remicade® (an anti-inflammatory).

Recombinant DNA technology also opened the doorway to produce mAbs with even lower murine composition. These humanized mAbs (-zumab), retaining murine sequence in the complementarity-determining region (CDR) only, were first produced in 1986 (*28*). As with chimeric technology, approval of the first humanized mAb therapeutic followed approximately 10 years later (Zenapax® for transplant rejection). A large number of humanized mAb products have since been successfully marketed, including Synagis®, Herceptin®, Mylotarg®, Xolair®, and Avastin®.

Chimeric and humanized antibody therapeutics are often produced in murine-derived cells. NS0 and SP2/0 myeloma cell lines, derived from B-lymphocytes of mice, have become commonplace for therapeutic development because they can be adapted to produce sufficiently high IgG titers in bioreactor cultures (*29*). NS0 cells, for example, lack the ability to express sufficient levels of glutathione synthase (GS), an enzyme necessary for biosynthesis of the essential nutrient glutamine. High-titer cell lines can be selected through co-transfection with a GS gene in a glutamine-free medium (*30*). Additional murine cells, such as CHO cells (derived from epithelial cells of Chinese hamster ovaries), have also become commonplace for drug development. CHO cells have the ability to produce self-sustaining levels of GS. However, GS inhibitors can be used in cell cultures to select only cells co-transfected with additional GS activity (*30*). Additional selectable markers, such as dihydrofolate reductase, can also be used for selection of suitably transfected clones (*20, 22, 30*). CHO cells as production hosts have been well-received by the biopharmaceutical community due to their ability to grow at high cell density and amenability to serum-free media (*23*). CHO cells have also been known for their production of proteins with a preferable glycoprofile, as described in more detail in following chapters: Mechanisms of Action chapter/Volume 1, Chapter 2 and Glycosylation chapter/Volume 2, Chapter 4. However, CHO cells have recently been reported to be capable of producing some of the undesired foreign glycan epitopes that are commonly produced in murine myeloma-based cell lines and were originally thought to be absent in CHO (e.g., gal-α-gal) (*31*). Throughout many years of development, a high level of process knowledge associated with NS0, SP2/0, and CHO has been compiled and will likely result in their continued use as platform cell lines for mAb production.

Considering the potential for murine epitopic determinants to elicit immunogenic responses, it makes sense that the production of fully human mAbs (-umab) for therapeutic use would also be explored. Transgenic mouse strains expressing human variable domains, phage display, and human-derived cell lines all offer the potential for fully human mAb expression (*32, 33*). Phage display, an *in vitro* technique that expresses and screens a library of antibody sequences, was the first technology to identify a fully human mAb for clinical use (*34*). The fully human construct for this mAb product (Humira®) was later transferred to a CHO cell expression system for commercial-scale production and licensed as a tumor necrosis factor (TNF) inhibitor useful for rheumatoid arthritis, Crohn's disease, and plaque psoriasis (*35*). Human-derived cell lines are a logical target for expression of therapeutics as they possess the biosynthetic pathways for human glycosylation and other post-translational modifications (PTMs), thereby minimizing the risks associated with anti-mAb immune responses. Fully human cell lines developed for biopharmaceutical production include the human embryonic kidney cell line (HEK 293) and its successors, as well as the Per.C6 cell line derived from human retinal cells (*23*). Per.C6 cell lines have been shown to offer several advantages, including very high titers and the ability to provide stable cell lines without selection agents (*29*). Per.C6 and HEK cell lines can be used for the expression of fully human mAbs (*36, 37*). However, full-length mAbs from these expression systems (Per.C6 or HEK) have yet to gain market approval.

Mammalian cell culture using the aforementioned cell lines has clearly dominated production of mAb therapeutics, in large part due to their ability to produce human-like form and function. Product development with CHO, NS0, and SP2/0 cell lines will undoubtedly continue to contribute novel therapeutics. Fully human expression systems will also likely increase in popularity, and it should be noted that recent advances in microbial expression systems may soon begin to play a role in this ever-expanding market (*38*). Since the initial inception of mAb therapeutics, a range of mammalian culture-derived mAb drug products have been approved by the FDA and are currently in use, as described in Table 2 (note that only full-length mAb and Fab therapeutics are listed) (*20, 39–41*). Murine, chimeric, humanized, and fully human mAbs of IgG1, 2, and 4 subclasses are in current clinical use today and have revolutionized modern medicine.

A variety of mAb-related therapies, such as Fab, Fc-fusion proteins, and antibody-drug conjugates (ADCs), have also been developed using mammalian cell culture (*39, 42*). Fab therapeutics are composed only of the antigen-binding subunit of the mAb and, therefore, do not have effector function capabilities (see the Mechanisms of Action chapter/Volume 1, Chapter 2). They also do not contain glycosylation and, therefore, have been expressed in bacterial cell culture (e.g., Lucentis®, approved for treatment of macular degeneration) or expressed as full-length mAbs and further truncated enzymatically (e.g., ReoPro®, approved for use as an antithrombotic agent) (*41, 43*).

Fc fusion proteins and ADCs harness mAb biochemical activity as a means for improving the pharmaceutical properties of an attached active pharmaceutical ingredient (API) or peptide. Fc-fusion proteins utilize the FcRn recycling pathway to improve half-life and pharmacokinetic properties (*42, 44*). Examples of approved Fc-fusion proteins include cytotoxic T-lymphocyte antigen (Orencia®) and TNF receptor type 2 (Enbrel®) for rheumatoid arthritis (*42*). Recently, there has been a great interest in Fc-fusion proteins with blood clotting factors to improve their half-life and reduce the frequency of injections for treatment of hematological disorders (*45, 46*).

ADCs are comprised of small-molecule APIs conjugated to full-length mAbs. ADCs harness the antigen-binding affinity and specificity of the mAb to deliver an API (e.g., chemotherapeutic agent) to a specific physiological location (*42, 47*). For example, Kadcyla® is a conjugate of trastuzimab and a microtubule antagonist. The mAb binds a target cancer cell expressing the selective HER-2 marker and provides localized drug targeting of an otherwise globally cytotoxic API (*47*).

A variety of additional mAb-based therapeutic strategies are also under development, including smaller single-chain fragment variable (scFv) antibodies, bi-specific antibodies with the ability to bind two separate epitopes, and multimer constructs of antigen-binding domains (*48–51*). Although scFcs, bi-specifics, and multimer constructs have yet to gain market approval in the United States, each of them is based upon critical recombinant mAb components and subject to the same production, regulatory, and characterization considerations described throughout this book.

Table 2. FDA-Approved Monoclonal Antibody (mAb) and Antigen-Binding Fragment (Fab) Therapeutics as of July 2014 *,† (20, 39–41)

Trade Name	Nonproprietary Name	Company	Target‡	Cell Line	Isotype	FDA Approval	Therapeutic Indications Approved by FDA
Orthoclone®K3®	Muromonab-CD3	Centocor Ortho Biotech (Johnson & Johnson)	CD3	Murine ascites	Murine IgG2a	1986	Transplantation rejection
ReoPro®	abciximab	Centocor Ortho Biotech (Janssen) and Eli Lilly	GPIIb/IIIa	SP2/0	Chimeric Fab	1994	High risk angioplasty
Zenapax®	daclizumab	Roche	CD25	NS0	Humanized IgG1	1997	Transplantation rejection
Herceptin®	trastuzumab	Genentech (Roche)	HER-2	CHO	Humanized IgG1κ	1998	Breast cancer, metastatic gastric or gastro-esophageal junction adenocarcinoma
Remicade®	infliximab	Centocor Ortho Biotech (Janssen)	TNF-α	SP2/0	Chimeric IgG1κ	1998	Crohns disease, ulcerative colitis, rheumatoid arthritis, ankylosing spondylitis, psoriatic arthritis, plaque psoriasis

Trade Name	Nonproprietary Name	Company	Target‡	Cell Line	Isotype	FDA Approval	Therapeutic Indications Approved by FDA
Simulect®	basiliximab	Novartis	CD25	SP2/0	Chimeric IgG1κ	1998	Transplantation rejection
Synagis®	palivizumab	MedImmune (AZ)	RSV F protein	NS0	Humanized IgG1κ	1998	Respiratory syncytial virus
Campath®	alemtuzumab	Millennium and Genzyme	CD52	CHO	Humanized IgG1κ	2001	B-cell chronic lymphocytic leukemia
Humira®	adalimumab	Abbott (Abbvie)	TNF-α	CHO	Human IgG1κ	2002	Rheumatoid arthritis, juvenile idiopathic arthritis, psoriatic arthritis, ankylosing spondylitis, Crohn's disease, plaque psoriasis
Zevalin®	ibritumomab tiuxetan	Biogen Idec	CD20	CHO	Murine IgG1κ	2002	Non-Hodgkin's lymphoma
Bexxar®	tositumomab and iodine-131 tositumomab	Corixa and GSK	CD20	Hybridoma	Murine IgG2aλ	2003	Non-Hodgkin's lymphoma
Xolair®	omalizumab	Genentech (Roche) and Novartis	IgE	CHO	Humanized IgG1κ	2003	Asthma

Continued on next page.

Table 2. (Continued). FDA-Approved Monoclonal Antibody (mAb) and Antigen-Binding Fragment (Fab) Therapeutics as of July 2014 *,† (20, 39–41)

Trade Name	Nonproprietary Name	Company	Target‡	Cell Line	Isotype	FDA Approval	Therapeutic Indications Approved by FDA
Avastin®	bevacizumab	Genentech (Roche)	VEGF	CHO	Humanized IgG1κ	2003	Metastatic colorectal cancer, non-small cell lung cancer, metastatic breast cancer, glioblastoma multiforme, metastatic renal cell carcinoma
Erbitux®	cetuximab	ImClone (Eli Lilly), Merck Serono and BMS	EGFR	SP2/0	Chimeric IgG1κ	2004	Head and neck cancer, colorectal cancer
Tysabri®	natalizumab	Biogen Idec and Elan	VLA-4	NS0	Humanized IgG4κ	2004	Multiple sclerosis (relapsing), Crohns disease
Lucentis®	ranibizumab	Genentech (Roche)	VEGF-A	E. Coli	Humanized Fab IgG1κ	2006	Macular degeneration and macular edema
Soliris®	eculizumab	Alexion Pharmaceutical	Complement C5	Myeloma	Humanized IgG2κ	2007	Paroxysmal nocturnal hemoglobinuria

Trade Name	Nonproprietary Name	Company	Target‡	Cell Line	Isotype	FDA Approval	Therapeutic Indications Approved by FDA
Cimzia®	certolizumab pegol	UCB	TNF-α	E. Coli	Humanized Fab IgG1κ	2008	Crohns disease, rheumatoid arthritis
Arzerra®	ofatumumab	Genmab and GSK	CD20	NS0	Human IgG1κ	2009	Chronic lymphocytic leukemia
Ilaris®	canakinumab	Novartis	IL-1β	SP2/0	Human IgG1κ	2009	Cryopyrin-associated periodic syndromes
Simponi®	golimumab	Centocor Ortho Biotech (Janssen)	TNF-α	SP2/0	human IgG1κ	2009	Rheumatoid arthritis, Psoriatic arthritis, ankylosing spondylitis
Stelara®	ustekinumab	Centocor Ortho Biotech (Janssen)	IL-12, IL-23	SP2/0	Human IgG1κ	2009	Plaque psoriasis
Actemra®	tocilizumab	Chugai (Roche)	IL-6	CHO	Humanized IgG1κ	2010	Rheumatoid arthritis

Continued on next page.

Table 2. (Continued). FDA-Approved Monoclonal Antibody (mAb) and Antigen-Binding Fragment (Fab) Therapeutics as of July 2014 *,† (20, 39–41)

Trade Name	Nonproprietary Name	Company	Target‡	Cell Line	Isotype	FDA Approval	Therapeutic Indications Approved by FDA
Prolia® and Xgeva®	denosumab	Amgen	RANKL	CHO	Human IgG2κ	2010	Postmenopausal osteoporosis, prevention of SREs in patients with bone metastases from solid tumors
Benlysta®	belimumab	HGS, GSK	BLyS	NS0	Human IgG1λ	2011	Systemic lupus erythematosus (SLE)
Yervoy®	ipilimumab	BMS	CTLA-4	CHO	Human IgG1κ	2011	Melanoma
Adcetris®	brentuximab	Seattle Genetics	CD30	CHO	Chimeric ADC IgG1κ	2011	Hodgkin lymphoma, systemic anaplastic large cell lymphoma
Perjeta®	pertuzumab	Genentech	HER2	CHO	Humanized IgG1κ	2012	HER2-positive metastatic breast cancer
Raxibacumab®	raxibacumab	HGS, GSK	PA of *B. anthracis* toxin	NS0	Human IgG1λ	2012	Anthrax exposure

Trade Name	Nonproprietary Name	Company	Target[‡]	Cell Line	Isotype	FDA Approval	Therapeutic Indications Approved by FDA
Gazyva®	obinutuzumab	Genentech	CD20	CHO	Humanized IgG1	2013	Chronic lymphocytic leukemia
Kadcyla®	ado-trastuzumab emtansine	Genentech	HER2	CHO	Humanized IgG1 ADC	2013	HER2-positive metastatic breast cancer
Cyramza®	ramucirumab	Eli Lilly and Co.	VEGFR2	NS0	Human IgG1	2014	Gastric cancer

* Only approved full-length mAb and Fab therapeutics are included. [†] Sources: *(20, 39–41)*. [‡] CD (cluster of differentiation), GPIIb/IIa (glycoprotein IIb/IIa), HER-2 (human epidermal growth factor receptor 2), TNF (tumor necrosis factor), RSV F protein (respiratory syncytial virus), IgE (immunoglobulin E), VEGF (vascular endothelial growth factor), EGFR (epidermal growth factor receptor), VLA-4 (very late antigen), IL (interluekin), RANKL (receptor activator of nuclear factor kappa-B ligand), BLyS (B-lymphocyte stimulator), PA of *B. anthracis* (protective antigen of *Bacillis anthracis*), CTLA (cytotoxic-lymphocyte antigen), VEGFR2 (vascular endothelial growth factor receptor 2).

Production of mAb Therapeutics

Current mAb biomanufacturing has evolved into a highly controlled process, as described in Figures 1 and 2. Each stage in the production process—raw materials, process conditions and control, purification, fill finish, and storage—can affect quality attributes of the product. The production process requires years of optimization and highly regulated control to result in a suitable drug product. At this point, it is useful to differentiate commonly used terminology associated with the drug development process (*52*). **Process-related impurities** refer to any unwanted material introduced as part of the manufacturing process. This may include impurities derived from the cell system itself (host cell proteins [HCPs] and DNA), cell culture media components, and impurities introduced during targeted purification strategies (e.g., column leachables, processing reagents). **Product-related impurities** are variants of the desired product (precursors, truncated products, or degradation products) that do not have the desired activity, efficacy, and/or safety. **Product-related substances**, on the other hand, are also variations of the targeted product; however, they fall within predefined specifications for activity, efficacy, and safety. A series of processing steps (upstream and downstream, described below) are undertaken to clear most unwanted process and product-related impurities to initially provide the bulk drug substance.

Bulk drug substance contains desired product as well as associated product-related substances and excipients/buffer components. The final stage in processing is formulation of the drug substance into a **drug product** suitable for clinical use. Formulation of the drug product may involve dilution to appropriate dosage and addition/removal of excipients into a pharmaceutical product for patient use. Drug product may have essentially the same identity and purity as drug substance other than it is in a format directly amenable for delivery to the patient. Therefore, although intended to be the same active ingredient, the storage conditions, shelf life, and degradation pathways may differ and should be thoroughly evaluated. In addition, specifications should be set for identity, purity (including any residual impurities), and potency, as described in ICH Q6B (*52*).

The development process is divided into upstream and downstream processing. Upstream process development involves cell line, media composition, and culture condition optimization to produce mAbs in sufficient quantity to support clinical and, subsequently, commercial production scale. A representative overview of an upstream production process is show in Figure 1, where the optimized cell line will undergo multiple rounds of expansion to first seed a small-scale bioreactor. These seed-scale expansion reactor cultures are transferred to one or more larger production reactors as necessary to produce sufficient levels of mAb. Early-stage purification steps, such as centrifugation and filtration, remove cellular debris and result in the clarified cell culture media (*53, 54*). Stable cell line development through delayed apoptosis, regulatory RNA, transient gene expression, improved cell culture media, single-use bioreactors, and process analytical technology (PAT) represent a sampling of recent advances in state-of-the-art upstream processes (*20, 22, 54–58*).

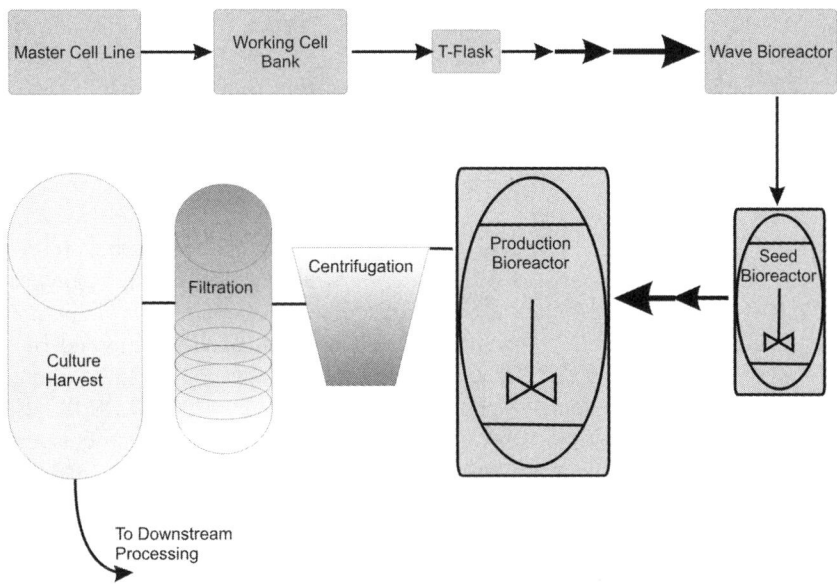

Figure 1. Representative upstream processing steps that may be used for monoclonal antibody production. (see color insert)

Following upstream processing, clarified cell culture media contains the desired mAb as well as other secreted HCPs, host cell DNA, media, feed components, and other potential process-related impurities. Downstream processing encompasses a variety of purification steps to selectively remove process-related impurities (Figure 2) (*59*). The first downstream processing stage is typically affinity enrichment of the IgG component of the culture media. This is most often achieved though protein A affinity chromatography, which selectively binds the Fc region. Although protein A is highly selective and can achieve purity greater than 90–98% (*60*, *61*), the inherent unit operation and limited binding capacity results in a relatively low throughput strategy (*62*). Optimization of ligand density, chromatographic support, immobilization strategy, and chromatographic parameters such as flow rate and buffer composition has been utilized to improve affinity enrichment capabilities (*63–65*) In addition, alternative ligands and non-chromatographic technologies have been explored to improve this initial capture step (*54*).

Despite the purity offered by protein A chromatography, residual impurities such as HCPs (Process Impurities chapter/Volume 2, Chapter 9 and LC-MS HCP chapter/Volume 3, Chapter 13) or adventitious agents (Adventitious chapter/Volume 3, Chapter 8) may remain after the initial capture step. HCPs may co-purify as adducts with the mAb of interest or as a result of nonspecific interaction and co-elution in the product fraction (*66*, *67*). Additional polishing chromatographic steps are often used such as cation exchange chromatography (CEX), anion exchange chromatography (AEX), and hydrophobic interaction chromatography (HIC) (*53*, *54*, *59*, *68*). A gel filtration step may also be present to remove aggregates during polishing chromatographic steps. The final stages of downstream process typically involved final filtration (nanometer-scale) and inactivation of potential viral contaminants, as well as ultrafiltration and/or dialysis to concentrate the product into its bulk drug substance form (*53*, *54*). Improvement in downstream processing is an ongoing area of research directed at achieving higher throughput purification to meet the demands of high-titer upstream production without sacrifice of drug substance purity. Many potential advances in chromatographic and non-chromatographic developments have recently been reviewed (*53*, *54*, *69–71*). Improvements in process-related technology continually are made as more sensitive and specific analytical technology for the detection and characterization of process-related impurities are developed. Emerging technologies for adventitious agent testing and HCP analysis are covered in detail throughout this book (Adventitious chapter/Volume 3, Chapter 8; Process Impurities chapter/Volume 2, Chapter 9; and LC-MS HCP chapter/Volume 3, Chapter 13). Genomics and proteomics have also bolstered the specificity of HCP identification and cell line-specific considerations (*72–74*), as described in chapters throughout this series (Genomics chapter/Volume 4 and Proteomics chapter/Volume 4).

Concurrent with the optimization of upstream and downstream processing to form a more pure and reproducible bulk drug substance, the material must undergo formulation development into a form suitable for direct clinical use. A variety of considerations go into drug product formulation, such as API concentration; dosage form (liquid vs. lyophilized); and selection of excipients and proposed storage conditions, including the container closure (e.g., vial, prefilled syringe). There is also increasing use of delivery devices such as auto-injectors and mini-dosers that allow for the delivery of high quantities of mAbs to patients.

The drug product matrix is of critical importance and ensures stability of the molecule throughout fill finish, transport, shelf life, and patient administration. Appropriate formulation minimizes chemical (e.g., proteolysis, disulfide scrambling, oxidation) and physical (e.g., denaturation, aggregation) instabilities and may include a variety of excipients such as carbohydrates, surfactants, polyols, and arginine or other amino acids (*75*). Assessment of the protein's stability begins early in the drug development lifecycle and often can be a determining factor in the developability (Developability chapter/Volume 2, Chapter 7) of a candidate mAb. A wide variety of analytical and biophysical techniques (Biophysical chapter/Volume 2, Chapter 6 and SMSLS chapter/Volume 3, Chapter 6) are used in such manufacturability studies.

The overall goal of process and formulation development is to produce a quality product suitable for its intended use. The quality of the drug substance or product is evaluated experimentally based on a variety of attributes determined to be critical to safety and efficacy (e.g., identity, potency, purity) (*52*). Critical quality attributes (CQAs) are physical, chemical, biological, or microbiological properties that must be maintained within a predefined limit, range, or distribution to ensure product quality (*76*). The identification of CQAs and evaluation of their level of criticality is a complex task that spans the totality of knowledge for a given process and product. A risk-based approach is taken to optimize and correlate all aspects of the production process for the severity of deviation from predefined specifications and the likelihood of each deviation. This combined approach of

CQA identification and correlation to process parameters followed by systematic process optimization is referred to as quality by design (QbD).

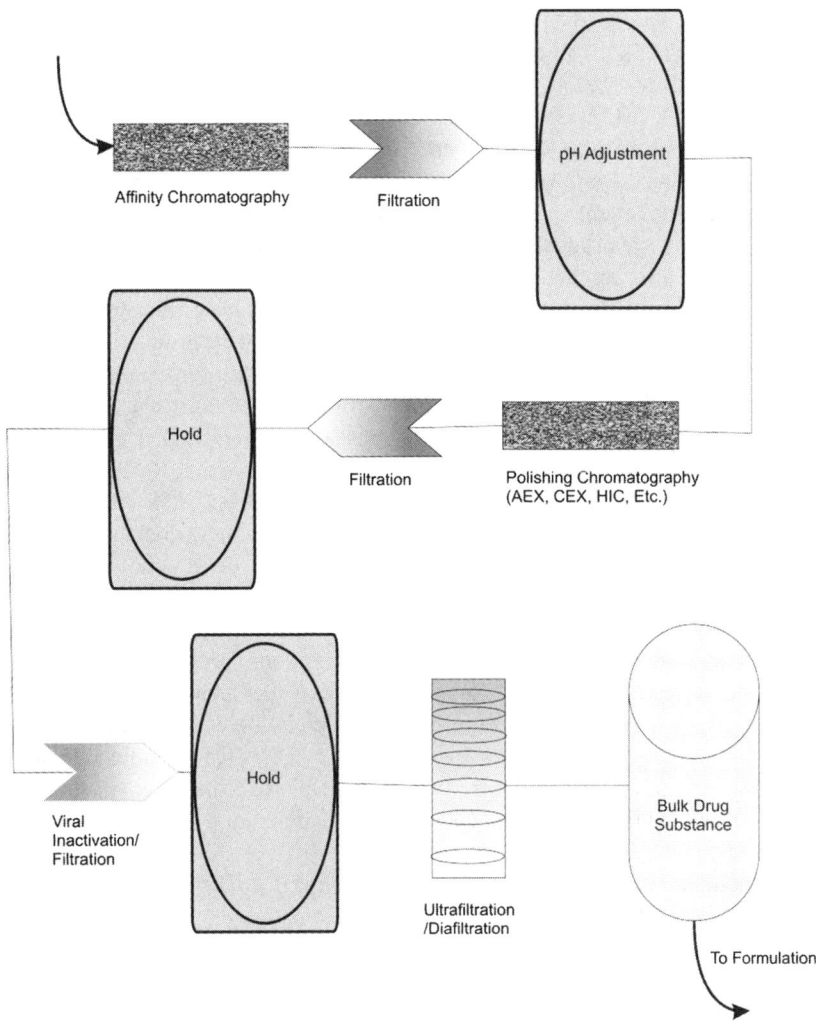

Figure 2. Representative downstream processing steps that may be used for monoclonal antibody production. Potential polishing chromatography steps include anion exchange chromatography (AEX), cation exchange chromatography (CEX), and hydrophobic interaction chromatography (HIC). (see color insert)

QbD is formally defined in ICH Q8(R2) as "a systematic approach to development that begins with predefined objectives and emphasizes product and process understanding and process control, based on sound science and quality risk assessment" (76). ICH Q8(R2) gives a generalized guidance on how the concept of QbD can be incorporated into pharmaceutical development. In 2008, the chemistry, manufacturing, and controls (CMC) biotechnology working group organized a comprehensive, real-world case study to more comprehensively exemplify all aspects of QbD principles based on a representative humanized IgG1, A-mAb (*77*). The case study used a subset of quality attributes (e.g., aggregation, glycosylation) known from historical knowledge to span a range of criticality. A risk assessment was described to demonstrate how historical, analytical, biophysical, clinical, and nonclinical data were combined for evaluation of A-mAb CQAs. The A-mAb case study then went on to present an iterative risk assessment and optimization strategy for upstream and downstream processing, linking product quality to critical control parameters. A similar exercise was also applied to drug product formulation design, leveraging historical process platform and product class information in combination with risk assessment. Finally, cumulative risk assessment along with product and process knowledge were used to define a control space and strategy for the representative QAs and CQAs that were further assured throughout the lifecycle with product and process verification (see the QbD chapter/Volume 1, Chapter 5 for a more detailed discussion).

PAT is a very important component of a robust QbD approach to biomanufacturing. The concept of PAT utilizes process and product knowledge to incorporate measurements (on-line, in-line, at-line, or off-line) of raw and in-process materials to provide real-time information as to the control of the system and ensure product quality (*76*, *78*). PAT measurements are intended to correlate critical process parameters and resultant product CQA's. The complexity and variability associated with protein therapeutics, raw materials, and their production process make information-rich PAT a difficult task; however, significant advancements have been realized (*79*, *80*). Cell culture operations are widely monitored for biomass (yield), critical reagents (e.g., metabolites, nutrients), and medium conditions (pO_2, pH, and temperature) through a variety of image analysis (e.g., focused beam reflectance), spectroscopic (e.g., IR, Raman), electrochemical (e.g., pH, dielectric spectroscopy), and/or off-line analytical techniques (e.g., high-performance liquid chromatography [HPLC], nuclear magnetic resonance [NMR]) (*78*, *79*). PAT has also been applied to harvest unit operations, downstream processing, and formulation (*77*, *79*, *81*). The vast array of available PAT tools has also spurred a movement toward multivariate statistical models for these complex data sets (*82*). Although a complete discussion of PAT is outside the scope of this chapter, many reviews and the A-mAb case study present the correlation between QbD, PAT and process control, and resultant product quality (*58*, *79*, *80*, *83*).

The A-mAb case study is a good example of widespread industrial collaboration to harmonize thinking and significantly advance antibody production philosophy and applied science. Although every aspect of A-mAb will not be directly applicable to every future mAb product, widely available case studies on representative materials are critical to advancing the science of complex mAb development in concert with regulatory requirements and expectations. It is the hope that the NISTmAb IgG1κ, described throughout this book, can serve a purpose similar to that of the A-mAb study, in this case, focusing on evaluation of current and future analytical and biophysical technology for identification and characterization of mAb product attributes. The NISTmAb will provide a common material to serve as a fundamental measurand of mAb heterogeneity, as demonstrated throughout this book.

Despite stringent controls and highly regulated manufacturing processes, the biological origin of recombinant therapeutics produces a significant level of product heterogeneity. Product-related substances consist in part of a variety of PTMs (PTMs chapter/Volume 2, Chapter 3), sequence variants (Sequence Variant chapter/Volume 2, Chapter 2), and other modifications that can be identified using techniques discussed throughout this book. In addition, the final product must be free of adventitious agents and have acceptable limits of product- and process-related impurities. Identification and control of these process variables and their effects on product quality is of great importance early in product development of mAb products to reduce costly development choices and influence early process decisions. Ultimately, it is the attributes of drug substance and drug product that determine its fitness for an intended use. Product safety and efficacy are initially verified through preclinical and clinical trials, and quality must be ensured thereafter through stringent analytical testing to ensure consistency from batch to batch. Process performance and product quality are tracked and trended over time to ensure product consistency. Changes in the production process are critically evaluated for resultant comparability to previous lots or reference standards using a full battery of characterization methods (*84*). These physicochemical and biophysical analytical technologies are used to "define" the product, as described in Volume 2 of this series, and many of these methods will support the validation of quality testing for lot release and stability. To ensure consistent production, it is therefore essential to have a reference standard of the specific product for comparison. Note that throughout this chapter, the words "in-house reference standard" are intended to refer to a company-specific product and "reference standard" alone refers to a standard issued by the World Health Organization (WHO) or a pharmacopoeial registry to assist in ensuring identity, potency, and/or purity. The term "reference material" refers to national metrology materials with metrological traceability, as discussed below.

Product-Specific In-House Reference Standards

Ultimately, it is the responsibility of the manufacturer to ensure product consistency throughout its lifecycle, using appropriate analytical characterization and comparability to an in-house reference standard (*52*). The drug candidate development process, appropriate in-house reference standards, and analytical methods co-evolve throughout the product lifecycle. In the case of mAbs, there currently is not a repository of product-specific compendia standards (described below) as those provided for small molecule drugs. Therefore, current best practices require development of a product and manufacturer-specific in-house reference standard. In-house reference standard development evolves as the product moves through various stages of clinical development as described in Figure 3. The timeline described in Figure 3 is a general outline of a theoretical situation, and the actual timeline for qualification of in-house reference standards and analytical methods is highly depend on real-time process and product knowledge as well as incoming data from clinical trials.

An **in-house interim reference standard** is an appropriately characterized lot of production material set aside for quality control (QC) purposes during the *development stage*. The interim standard is often used as the product-specific reference standard for early technical development through Good Laboratory Practice-Toxicology (GLP-Tox) and early clinical studies. At this point, tentative process parameters and formulation for Phase 1 clinical trials have been defined, and a suite of analytical characterization methods to define product properties such as primary sequence, certain PTM modifications, charge and size isoforms, and potency are in place. These methods are used to **qualify the reference standard**, which refers to collection of sufficient physicochemical and biophysical characterization data such that the material can serve as a representative comparator for future lots and analytical method evaluation. During Phase 1 and Phase 2 trials, process changes may occur. The need to replace the interim standard will depend on the level of process change and/or detected changes to the product profile post-change. Qualification of new interim standards should be minimized to avoid product drift, but the decision must ensure the in-house reference standard is representative of the product to be used in the clinic so that it is fit for its intended purpose. When a significant process change is to be implemented that impacts relevant quality attributes, a batch of designated in-house reference standard should be simultaneously generated and characterized, the data from which can also be used as the basis for a comparability exercise (*84*).

At or near pivotal clinical trials, the overall upstream, downstream, and formulation scheme intended for commercial development will be in place. The entire suite of analytical characterization and QC (lot release and stability) methods should now be qualified and validated, respectively, as discussed in the following paragraph. A larger quantity of manufacturer's material must be selected from a batch that is representative of the commercial product for use in pivotal trials and post-commercialization. This batch is often split into two subsets for use as an **in-house primary reference** standard and the first lot of **in-house secondary (or working) reference standard**. The in-house primary standard is expected to be in quantities sufficient to be used throughout the product lifecycle for qualification/calibration of secondary standards. The **in-house secondary reference standard** is calibrated against the in-house primary reference standard and is used in QC testing of clinical material as well as marketed lots. Additional batches of secondary in-house reference standard may be made when supplies are exhausted and re-qualified against the in-house primary reference standard. Additional in-house primary reference standard may also be prepared if the initial batch is near exhaustion or changes in the reference profile are noted during regularly scheduled trending testing. However, qualification of new in-house primary reference standard should be avoided when possible to minimize potential drift.

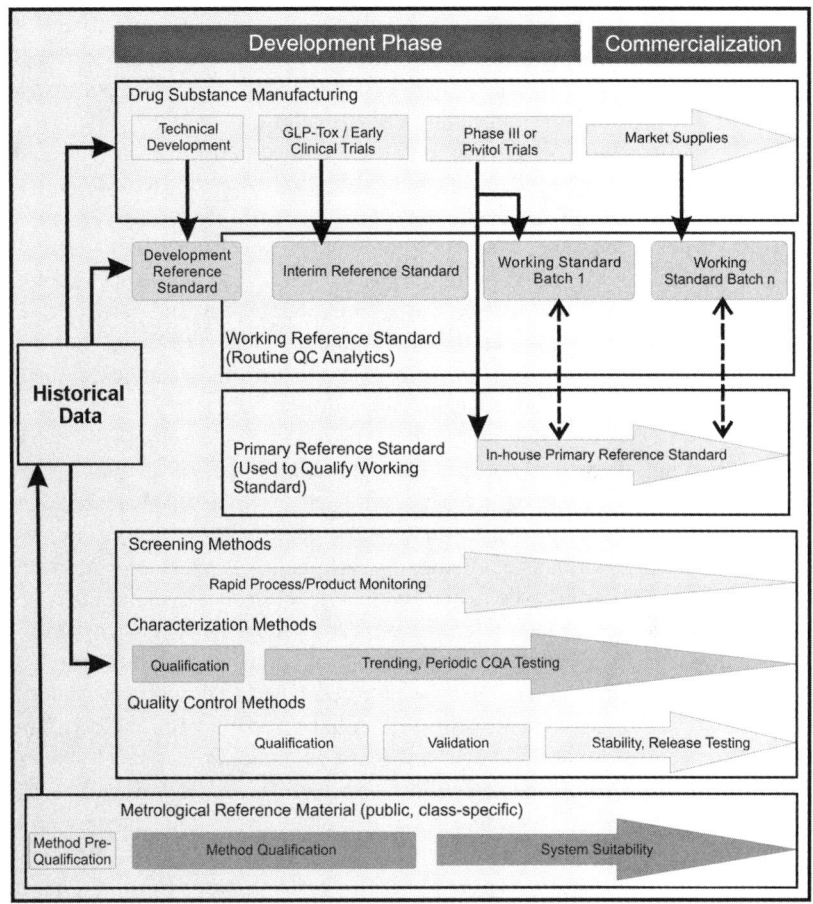

Figure 3. Representative monoclonal antibody lifecycle incorporating potential timelines for analytical method development, in-house reference standards, and potential supplementation with a metrological reference material. (see color insert)

Along with evolution of in-house reference standards and the drug development lifecycle, analytical methods for product testing also evolve in a manner appropriate to the current stage of development (*85*). **Method Qualification** refers to the use of an in-house standard along with challenge material (e.g., forced degraded material, known impurities) to test the ability of a method to provide information on the desired product attribute. For example, a method qualified for identity testing should be sufficient to differentiate the test subject from product-related impurities and other related molecules produced in that facility. **Method Validation** is a more in-depth verification of a proposed method's suitability for an intended purpose, as described in more detail in ICH Q2(R1) (*86*). Succinctly, validation consists of a method performance evaluation for accuracy, precision, specificity, detection limits, linearity, and range to provide a high degree of assurance that it is capable of consistently producing results within predetermined specifications for a given product. Analytical and biophysical methods may categorized as being informational (i.e. for research purposes), qualified, and/or validated, depending on their intended purpose and current role in a particular drug candidate's lifecycle.

Analytical and biophysical methods can be split into three categories that co-evolve throughout the product lifecycle. The first types of methods typically utilized during early product development are methods used to screen candidate molecules. These techniques assess for commonly known undesirable attributes such as a significant propensity to aggregate or high levels of product variability (e.g., in size, charge, or viscosity). *In vitro* immunogenicity and potency assays also play a significant role in determining viable candidates to move forward in development. In addition to these screening assays, promising candidates may be subjected to further, more detailed characterization.

Detailed characterization methods are used for high-level product understanding and often focus on specific product attributes of identity (primary sequence and higher order structure), and purity, which is evaluated based on intrinsic heterogeneity such as PTMs, sequence variants, size variants (aggregation), charge variants, and other characteristics. Product characterization methods have an important role as the product development cycle advances and may include mass spectrometry (e.g., for sequence determination, glycoprofiling), HPLC (along with fraction collection of size, charge, and sequence variants), and methods that focus on higher order structure (e.g., NMR, hydrogen-deuterium exchange [HDX], circular dichroism, differential scanning calorimetry). These methods are generally not validated, but must be qualified to a level shown to be fit for use when compared to routine lot release methods intended for QC. This is because they yield critical data on potential product changes during process optimization, such as changes in primary sequence; PTMs; biophysical parameters; and secondary, tertiary, and higher order structure. Qualified characterization methods are often used to supplement the application for licensure (in the elucidation of structure section in the application), demonstrate a high level of product knowledge, and verify that more robust QC methods are fit for their intended purpose. Qualified characterization methods are required for qualification of future lots of in-house working standards. Trending characterization data associated with sequential production lots is also critical to post-approval comparability exercises that may be necessary to justify process changes have not adversely impacted the product (*84*).

A more robust set of methods with defined precision and acceptance criteria (QC methods) must also be developed for ensuring the product quality and stability of future clinical and commercial lots. Both screening and characterization methods may eventually become QC methods, depending on their suitability for an intended purpose. Characterization methods are often used to assess and/or supplement intended QC methods because of their ability to accurately and precisely identify deviations from desired quality attributes. Qualification of QC methods begins during early toxicology studies and clinical trials. As manufacturing processes are scaled to levels required to support pivotal clinical trials, multiple lots of material are used to validate such assays for their intended purpose. Appropriate validation of an analytical method serves to confirm acceptance criteria and suitable performance, as defined in earlier qualification studies. Validated QC methods are used to assure consistent production of commercial lots. In-house working standards and controls are run alongside commercial material to ensure QC method conformance to expectations and, ultimately, a safe and effective product for clinical and commercial use.

Finally, robust operation of characterization and QC methods is balanced by continued evaluation of system performance. All components, including instrumentation, consumables, software, and analytical personnel, should be included in the system suitability space of a particular test method. Historically, system suitability has been established through use of an in-house reference standard. Conformance to expectation indicates proper operation of a test system. Although consistent operation of a test system that has been validated to be capable of identifying a particular product change is strong evidence for product consistency, when deviations from the expected outcome are observed, additional mechanisms are needed to differentiate product versus method-related factors. For this reason, external, non-product-specific standards or reference materials are necessary to challenge analytical operations and yield a secondary confirmation of system suitability.

Metrological Reference Material

Establishing appropriate public reference standards for pharmaceutical development is a collaborative effort involving drug manufacturers, regulatory agencies, and a variety of standards organizations with unique yet overlapping missions. The WHO coordinates the development of standards associated with health care throughout the United Nations system. The WHO provides international reference materials (International Biological Reference Preparations) useful for designating a baseline definition of unit activity in a particular assay (e.g., potency assay) (*87*). These materials are intended for calibration of national/regional activity reference standards (e.g., United States Pharmacopeia) and/or in-house reference standards with regards to product potency or biological activity. These potency standards represent the gold standard for activity. However, for new molecular entities and all currently approved mAbs, such standards are typically not available. In this case, qualification and definition of activity are based on a representative lot of an in-house reference standard, as described above. In the case of a follow-on biological program, the originator molecule must be purchased from market supplies and used to assess an in-house standard manufactured by the follow-on manufacturer.

In the United States, the U.S. Pharmacopoeial Convention (USP) publishes official compendia for pharmaceutical products, the content of which are often enforceable by the FDA (*88*). The USP was established with the mindset that drug substances and products are articles of commerce that must pass stringent quality standards to prevent distribution of adulterated or misbranded products, thereby ensuring safety and efficacy for clinical use (*89*). To this end, the USP publishes product-specific pharmacopoeia monographs, including methodology and appropriate calibrants to aid in assessing whether a product meets required specifications. Requirements for defining identity, purity, stability, and potency for small molecule drugs are well-established due to the definitive chemical structure of such molecules (*88*). It is the intent of the USP that every legally approved article (e.g., drug substance, drug product) should have a monograph and a USP Reference Standard, where appropriate. To this end, the USP works diligently with originator manufacturers to develop monographs and associated reference standards. Alternative sources such as potential generic suppliers may also be sought as sources, however, if no monograph and reference standard are under development 5 years prior to the expiration of an originator patent (*90*). This mechanism has worked well for small molecule drugs, for which compendial methods and standards are typically available at or near the time of patent expiry.

In the case of biologically derived medicines, their inherent complexity requires additional consideration for attributes such as identity, stability, product-related impurities, and process-related impurities. To date, no mAb monographs have been published in the legally enforceable USP compendium. However, a monograph for rituximab is available through the non-mandatory USP Medicines Compendium (*91*). Although not legally enforceable unless submitted as part of a regulatory filing, Medicines Compendium standards are approved through extensive USP Expert Committee evaluation and may be useful to establish an article's identity, strength, and purity. In addition to product-specific monographs, the USP publishes General Chapters (often with associated procedural standards) aimed at best practices for techniques that may be broadly applied to a variety of health care-related products, including the future inclusion of a recently created chapter on size, charge, and glycosylation testing for mAbs (*92*). Other country-based or regional pharmacopoeial agencies are also publishing standards related to mAb drug substances and products. The Indian Pharmacopoeia is in advanced stages of publishing monographs for rituximab drug substance and drug product for injection (*93*). The European Pharmacopoeia includes a general monograph titled *Monoclonal Antibodies for Human Use* (2031), which provides definitions and general provisions for production, testing, and labeling. It is likely that major pharmacopoeial agencies in Japan, China, Brazil, and other countries will follow suit.

During method development and in-house reference standard evolution of complex drug products such as mAbs, it makes sense that the best comparability standard is a representative lot of the specific drug substance or product itself (in-house primary and working standards). In some cases, biologic pharmacopoeial standards such as erythropoietin (EPO) and granulocyte colony stimulating factor (GCSF) are available from Pharmacopoeia. However, physicochemical and biophysical standards of this type are typically not available for biotherapeutics. Given the process-specific nature and high complexity of mAb products, it may be impossible for one national or international reference standard to cover all of the needs when testing a company-specific product. Therefore, multiple company-specific lots and often attribute-specific reference standards (e.g., certain degradation products derived thereof) will be required to ensure a method's performance for a particular biopharmaceutical product, and an in-house reference standard will be required to rigorously monitor product consistency. The necessity for method validation and guidance for such an endeavor has been stated by a variety of regulatory and standardization organizations (*52, 86, 94*), and many excellent reviews have recently been published (*95, 96*). It has been noted that guidelines are subject to some level of user interpretation, which can lead to inadvertent risk if appropriate validation parameters are not considered (*96*). However, the interpretability of guidance documents is also an essential factor that allows consideration of the totality of evidence for a specific product. A widely available metrological reference material would provide a representative material to more precisely define a balance between harmonization and product-specific validation packages. In addition, appropriate protocols for method qualification during early- to mid-phase product development are not as harmonized or clearly defined as those for later phases of product development because most regulatory guidance documents are designed for commercialization of a product (*85, 95*). Such a void in qualification and assessment of changing analytical test methods would, therefore, be supplemented by a widely available metrological reference material and reference data to supplement current in-house reference standard protocols.

National metrology institutes such as NIST are responsible for such metrological reference materials as one aspect of assuring measurement equivalence. These institutes are involved in a variety of activities, ranging from establishing the fundamental unit of time measurement to providing physical reference materials useful for calibrating property measurements such as mass. To achieve this mission, a national metrology institute may provide chemical and physical reference materials to its stakeholders to establish a route of traceability to fundamental measurement units and/or assess the quality of a measurement procedure. In the health care setting, reference materials are often used for calibration and/or harmonization of test methods and focus on the accuracy and reproducibility of measurement technology itself, as opposed to assessing a specific product's conformance to predefined specifications. Metrological reference materials such as the NISTmAb described here are, therefore, similar to procedural standards established in USP compendia, and are intended to compliment these activities by providing a widely representative and internationally traceable material for analytical method assessment.

The NIST Biomanufacturing Program is directed toward developing a suite of fundamental measurement science, reference materials, and reference data to enable more accurate and confident characterization of key attributes directly linked to product safety and efficacy. A critical metric in achieving these goals is the production of a widely available reference material useful for establishing instrument performance and variability in analytical test methods (*97*). Recombinant mAbs are the fastest growing class of biotherapeutics and are, therefore, an obvious candidate for such a material. A **NIST reference material (RM)** is a material that is sufficiently homogeneous and stable with reference to specified properties and has been established to be fit for its intended use in measurement or in examination of nominal properties. The topic of the current book is a candidate IgG1κ mAb RM for which detailed analytical and biophysical characterization will be presented. Property values of an RM are a best estimate of the true value provided by NIST where all known or suspected sources of bias may not have been fully investigated. NIST RMs meet the International Organization for Standardization (ISO) definition of a reference material, including homogeneity, stability, and suitability for use in a measurement process (98). A **NIST Standard Reference Material (SRM)** is a material accompanied by documentation issued by NIST that assigns one or more specified property values with associated uncertainties and traceability. Property values of an SRM are certified as being traceable to an accurate realization of the unit in which the property values are expressed and having suspected sources of bias that have been fully investigated or accounted for by NIST. NIST SRMs meet the ISO definition of a certified reference material (CRM) (98). Both NIST RMs and SRMs are issued under the NIST trademark and can be used for measurement quality assurance.

The subject IgG1κ discussed throughout this book is intended for development into an RM and/or SRM that is expected to be used by a variety of stakeholders, including the biopharmaceutical industry, instrument manufacturers, academia, regulatory authorities, and other standards organizations. The RM is intended for a variety of uses, including, but not necessarily limited to, system suitability tests, establishing method or instrument performance and variability, comparing changing analytical test methods, and assisting in method qualification. To properly serve as a quantitative and qualitative RM, a variety of physical and chemical characterization methods may be used to determine biomolecular composition and structure, purity, and stability, including, but not necessarily limited to, liquid chromatographic methods; mass spectrometry; NMR; and optical, X-ray, and other product characterization assays.

Information pertaining to chemical and physical attributes of the NISTmAb RM or SRM may be reported to customers as NIST Certified Values, NIST Reference Values, or NIST Informational Values, depending on the level of certainty associated with the particular test methods. Analytical data may also be made available in a variety of formats. including, but not necessarily limited to certification sheets delivered with the material, an SRM or RM website, Standard Reference Data software and/or databases, or published material in scientific journals and books such as the current series. Characterization efforts throughout this book utilized the candidate RM 8670 (lot 3F1b) of the NIST IgG1κ mAb (100 mg/mL or 10 mg/mL). The molecule was distributed throughout industry, academia, regulatory agencies, and NIST to gain initial product understanding and identification of its physicochemical and biophysical attributes. The intention was to evoke best practices in a collaborative effort toward characterization of a mAb. Simultaneously, additional NISTmAb material intended for public release as an RM and/or SRM was prepared from multiple homogenized production lots and is expected to be available shortly after publication of this series.

Potential Utility of the NISTmAb IgG1κ

The pursuit of a candidate NIST RM is based on a variety of factors that stem from industry input. The decision to pursue an IgG1κ mAb RM arose largely through discussions and iterative research with industry stakeholders over a period of 5 years. mAbs of a given class are highly homologous and, therefore, have similar characteristics for which platform technology can provide a wealth of information. Screening methodologies for class-specific attributes are commonly developed with this highly similar behavior and composition in mind. However, start-up companies may not have such historical expertise, and/or investigational compounds may be present in very short supply. One of the strengths of having an established RM of the IgG1κ class is to assist with development and optimization of such techniques for new molecular entities. The NISTmAb reference material is expected to fill this void as a representative material for method prequalification during early drug development, as depicted in Figure 3, and feed forward into class-specific historical knowledge. One could imagine a series of follow-on isotypes, allotypes, or other class-specific molecules to support development of a variety of therapeutic proteins.

Further method development, incorporation of novel analytical and biophysical techniques, or method transfer (internal or to a contract organization) also requires a high level of analyte knowledge to evaluate suitability. Instrument vendors and industry consumers alike often use company-specific mAbs with intellectual property concerns, commercially available mAbs that may not be well-characterized, or proteins not representative of the class for such a purpose. The use of a single available material will be convenient for users and instrument developers alike to evaluate the instrument or method performance of evolving technology. Certified concentration and extensive characterization data collected by multiple companies and/or institutions provided along with such a material will greatly facilitate determination of dynamic range, detection limits, linearity, and precision of new technology. Again, although the use of degraded material or other products produced in the same facility is required for challenging methods, the NISTmAb will provide an external control that can be widely utilized to evaluate purity or identity-indicating assays. The baseline comparator NISTmAb molecule will, therefore, facilitate implementation of new characterization and/or QC strategies.

In addition, the historical data available for direct comparison will assist regulators in evaluating the suitability of new techniques for use in originator product licensure applications. The inevitable submission for follow-on biologic licensure is an even more pressing issue due to the expected impact of increased analytics and reduced clinical trials. Every aspect from sample handling to instrument performance must be verified to ensure precise and accurate method readouts. Technology associated with a follow-on antibody submission may differ greatly from legacy methods utilized for the originator product. Regulatory officials and developers therefore would greatly benefit from a goalpost molecule that can differentiate method-related artifacts from those inherent to the product and/or claims of similarity from multiple follow-on submissions.

The entire biopharmaceutical design space depicted in Figure 3 relies heavily on historical knowledge, including previous discovery platforms, cell line and process knowledge, appropriate production and use of in-house reference standards, and the analytical and biophysical expertise required to characterize such standards. The metrological IgG1κ reference material is intended to provide a widely available test product that is not associated with product-specific intellectual property concerns. Historical data and widespread availability of such a material will be useful for a broad community assessment of current and emerging analytical technology and will establish a more robust framework for method qualification. Historical product knowledge associated with the RM may serve to feed forward into the drug development process, thereby allowing more informed selection of test methods appropriate for mAb products and supplementing the totality of evidence that a specific method is capable of producing results in accord with its intended purpose.

Concluding Remarks

The development of mAb therapeutics is an astounding story of how groundbreaking research can translate into viable lifesaving products. In less than 30 years, significant biochemical discoveries have now resulted in novel treatments for numerous indications that have had an invaluable impact on patients worldwide. Continued collaboration between academia, industry, and federal agencies (as evidenced by the current collaborative series) demonstrates that this trend in innovative mAb health care will continue for years to come. As of April 2014, there were 30 mAb therapeutics in Phase 3 clinical trials (99). The proven mAb therapeutic track record as a sustainable and necessary health care market warrants addition of metrological standards and establishment of best practices for characterization.

The metrological reference material will not replace in-house reference standards, but rather will supplement best practices historically used to ensure product quality. The current project represents two very important milestones in furthering development of monoclonal therapeutics. The NISTmAb will first be subjected to state-of-the-art characterization practices as determined through a large interagency collaborative effort, setting a benchmark for mAb characterization and a forward-looking presentation of next-generation analytical methods. Simultaneously, historical data is being generated on this reference material similar to what typically would be performed on a primary in-house reference standard. This material is beginning its journey through a mAb lifecycle, and will serve as a tangible, openly available substance to critically evaluate analytical questions related to product characterization, method development, and in-house reference standard programs. Although the establishment of a suitable reference material for complex mAbs comes with qualitative and quantitative analytical challenges that have not been faced previously, implementation will supplement the unrivaled commitment to biopharmaceutical quality demonstrated by analytical scientists to improve the safety and efficacy of biopharmaceuticals.

Disclaimer

Commercial equipment, instruments, and materials are identified in this paper to adequately exemplify the discussion and experimental procedure. Such identification does not imply recommendations or endorsements by NIST nor does it imply that the equipment, instruments, or materials are necessarily the best available for the purpose.

References

1. Wachtel-Galor, S; Benzi, I. *Herbal Medicine: Biomolecular and Clinical Aspects*; CRC Press: Boca Raton, FL, 2011.
2. Jones, A. W. Early drug discovery and the rise of pharmaceutical chemistry. *Drug Test. Anal.* **2011**, *3*, 337–44.
3. Kumar, H.; Kawai, T.; Akira, S. Pathogen recognition by the innate immune system. *Int. Rev. Immunol.* **2011**, *30*, 16–34.
4. Goldsby, R.; Kindt, T.; Osborne, B.; Kuby, J. *Immunology*, 5th ed.; W.H. Freeman and Company: New York, 2003.
5. Deng, L.; Luo, M.; Velikovsky, A.; Mariuzza, R. Structural insights into the evolution of the adaptive immune system. *Annu. Rev. Biophys.* **2013**, *42*, 191–215.
6. Parra, D.; Takizawa, F.; Sunyer, J. O. Evolution of B cell immunity. *Annu. Rev. Anim. Biosci.* **2013**, *1*, 65–97.
7. Von Behring, E.; Kitasato, S. Uber das zustandekommen der diphtherie-immunitat und dr tetanus-immunitat bei tieren. *Dtsch. Med. Wochenschr* **1890**, *16*, 1113–1114.
8. Cenci, F. Alcune experienze di sieroimmunizzaziuone e sieroterapie nel norbillo. *Riv. Clin. Pediatr.* **1907**, *5*, 1017–1025.
9. Cohn, E. Blood proteins and their therapeutic value. *Science* **1945**, *101*, 51–56.
10. Cohn, E. J.; Strong, L. E.; Hughes, W. L.; Mulford, D. J.; Ashworth, J. N.; Melin, M. Preparation and Properties of Serum. *J. Am. Chem. Soc.* **1946**, *68*, 459–475.
11. Eibl, M. M. History of immunoglobulin replacement. *Immunol. Allergy Clin. North Am.* **2008**, *28*, 737–64, viii.
12. Kohner, G.; Millstein, C. Continuous cultures of fused cells secrieting antibody of predefined specificity. *Nature* **1975**, *256*, 495–497.
13. *General Policies for Monoclonal Antibodies*; INN Working Document 09.251; World Health Organization: Geneva, Switzerland, 2009.
14. *International Nonproprietary Names (INN) for Biological and Biotechnological Substances (A Review)*; INN Working Document 05.179; World Health Organization: Geneva, Switzerland, 2013.
15. Van den Hoogen, M. W. F.; Hilbrands, L. B. Use of monoclonal antibodies in renal transplantation. *Immunotherapy* **2011**, *3*, 871–80.
16. Smith, S. L. Ten years of Orthoclone OKT3. *J. Transpl. Coord.* **1996**, *6*, 109–121.
17. Jones, S. D.; Castillo, F. J.; Levine, H. L. Advances in the development of therapeutic monoclonal antibodies. *BioPharm Int.* **2007**, 96–114.
18. Delmonico, F.; Fuller, T.; Russell, P. Variation in patient response associated with different preparations of murine monoclonal antibody therapy. *Transplantation* **1989**, *47*, 92–95.
19. Caron, P. C.; Schwartz, M. a; Co, M. S.; Queen, C.; Finn, R. D.; Graham, M. C.; Divgi, C. R.; Larson, S. M.; Scheinberg, D. a. Murine and humanized constructs of monoclonal antibody M195 (anti-CD33) for the therapy of acute myelogenous leukemia. *Cancer* **1994**, *73*, 1049–56.

20. Li, F.; Vijayasankaran, N.; Shen, A. (Yijuan); Kiss, R.; Amanullah, A. Cell culture processes for monoclonal antibody production. *mAbs* **2010**, *2*, 466–479.
21. Neslon, D.; Cox, M. *Principles of Biochemistry*, 4th ed.; W.H. Freeman and Company: New York, 2005; pp 307–317.
22. Cacciatore, J. J.; Chasin, L. a; Leonard, E. F. Gene amplification and vector engineering to achieve rapid and high-level therapeutic protein production using the Dhfr-based CHO cell selection system. *Biotechnol. Adv.* **2010**, *28*, 673–81.
23. Bandaranayake, A. D.; Almo, S. C. Recent advances in mammalian protein production. *FEBS Lett.* **2014**, *588*, 253–60.
24. Morrison, S. L.; Johnson, M. J.; Herzenberg, L. a; Oi, V. T. Chimeric human antibody molecules: mouse antigen-binding domains with human constant region domains. *Proc. Natl. Acad. Sci. U.S.A.* **1984**, *81*, 6851–6855.
25. Knight, D. M.; Trinh, H.; Le, J.; Siegel, S.; Shealy, D.; McDonough, M.; Scallon, B.; Moore, M. A.; Vilcek, J.; Daddona, P. Construction and initial characterization of a mouse-human chimeric anti-TNF antibody. *Mol. Immunol.* **1993**, *30*, 1443–1453.
26. The EPIC investigators. Use of a monoclonal antibody directed against the platelet glycoprotein IIb/IIIa receptor in high-risk coronary angioplasty. *N. Engl. J. Med.* **1994**, *330*, 956–961.
27. Knight, D. M.; Wagner, C.; Jordan, R.; McAleer, M. F.; DeRita, R.; Fass, D. N.; Coller, B. S.; Weisman, H. F.; Ghrayeb, J. The immunogenicity of the 7E3 murine monoclonal Fab antibody fragment variable region is dramatically reduced in humans by substitution of human for murine constant regions. *Mol. Immunol.* **1995**, *32*, 1271–1281.
28. Jones, P.; Dear, P.; Foote, J.; Neuberger, M.; Winter, G. Replacing the complementarity-determining regions in a human antibody with those from a mouse. *Nature* **1986**, *321*, 522–525.
29. Coco-Martin, J.; Harmsen, M. A review of therapeutic protein expression by mammalian cells. *Bioprocess Int.* **2008**, 28–33.
30. Birch, J. R.; Racher, A. J. Antibody production. *Adv. Drug Delivery Rev.* **2006**, *58*, 671–685.
31. Bosques, C.; Collins, B.; Meador, J.; Sarvaiya, H.; Murphy, J.; DelloRusso, G.; Bulik, D.; Hus, I.; Washburn, N.; Sipset, S.; Myette, J.; Raman, R. Chinese hamster ovary cells can produce galactose-a-1,3-galactose. *Nature* **2010**, *28*, 1153–1156.
32. Lonberg, N.; Taylor, L.; Harding, F. Antigen-specific human antibodies from mice comprising four distinct genetic modifications. *Nature* **1994**, *368*, 856–859.
33. McCafferty, J.; Griffiths, A.; Winter, G.; Chiswell, D. Phage antibodies: Filamentous phage displaying antibody variable domains. *Nature* **1990**, *348*, 522–554.
34. Nelson, A. L.; Dhimolea, E.; Reichert, J. M. Development trends for human monoclonal antibody therapeutics. *Nat. Rev. Drug Discovery* **2010**, *9*, 767–74.

35. Mease, P. J. Adalimumab in the treatment of arthritis. *Ther. Clin. Risk Manage.* **2007**, *3*, 133–148.
36. Yallop, C.; Crowley, J. PER.C6® Cells for the Manufacture of Biopharmaceutical Proteins. In *Modern Biopharmaceuticals: Design, Development and Optimization*; Knablein, J., Ed.; Wiley: Weinheim, 2008; pp 779–808.
37. Zhu, J. Mammalian cell protein expression for biopharmaceutical production. *Biotechnol. Adv.* **2012**, *30*, 1158–70.
38. Spadiut, O.; Capone, S.; Krainer, F.; Glieder, A.; Herwig, C. Microbials for the production of monoclonal antibodies and antibody fragments. *Trends Biotechnol.* **2014**, *32*, 54–60.
39. Ghaderi, D.; Zhang, M.; Hurtado-Ziola, N.; Varki, A. Production platforms for biotherapeutic glycoproteins. Occurrence, impact, and challenges of non-human sialylation. *Biotechnol. Genet. Eng. Rev.* **2012**, *28*, 147–176.
40. Spada, S.; Walsh, G. *Directory of Approved Biopharmaceuticals*; CRC Press: Boca Raton, FL, 2004.
41. E-mail: Drugs@FDA.com.
42. Rajpal, A.; Strop, P.; Yeung, Y.; Chaparro, J.; Pons, J. Introduction: Antibody structure and function. In *Therapeutic Fc-Fusion Proteins*; Wiley, New York, 2014; pp 1–28.
43. Nelson, A. L. Antibody fragments: Hope and hype. *mAbs* **2010**, *2*, 77–83.
44. Wu, B.; Sun, Y.-N. Pharmacokinetics of Peptide-Fc fusion proteins. *J. Pharm. Sci.* **2014**, *103*, 53–64.
45. Shapiro, A. Development of long-acting recombinant FVIII and FIX Fc fusion proteins for the management of hemophilia. *Expert Opin. Biol. Ther.* **2013**, *13*, 1287–97.
46. Kaufman, R.; Powell, J. Molecular approaches for improved clotting factors for hemophilia. *Blood* **2013**, *122*, 3568–3574.
47. Sliwkowski, M. X.; Mellman, I. Antibody therapeutics in cancer. *Science* **2013**, *341*, 1192–8.
48. Byrne, H.; Conroy, P. J.; Whisstock, J. C.; O'Kennedy, R. J. A tale of two specificities: Bispecific antibodies for therapeutic and diagnostic applications. *Trends Biotechnol.* **2013**, *31*, 621–32.
49. Frankel, S. R.; Baeuerle, P. A. Targeting T cells to tumor cells using bispecific antibodies. *Curr. Opin. Chem. Biol.* **2013**, *17*, 385–392.
50. Ahmad, Z. A.; Yeap, S. K.; Ali, A. M.; Ho, W. Y.; Alitheen, N. B. M.; Hamid, M. scFv antibody: Principles and clinical application. *Clin. Dev. Immunol.* **2012**, *2012*, 1–15.
51. Hess, C.; Venetz, D.; Neri, D. Emerging classes of armed antibody therapeutics against cancer. *MedChemComm* **2014**, *5*, 408–431.
52. *ICH Topic Q 6 B. Specifications, Test Procedures and Acceptance Criteria for Biotechnological Products*. International Conference on Harmonisation of Technical Requirements for Registration of Pharmaceuticals for Human Use (ICH), 1999.
53. Marichal-Gallardo, P. A; Alvarez, M. M. State-of-the-art in downstream processing of monoclonal antibodies: Process trends in design and validation. *Biotechnol. Prog.* **2012**, *28*, 899–916.

54. Shukla, A. a; Thömmes, J. Recent advances in large-scale production of monoclonal antibodies and related proteins. *Trends Biotechnol.* **2010**, *28*, 253–61.
55. Butler, M.; Meneses-Acosta, a. Recent advances in technology supporting biopharmaceutical production from mammalian cells. *Appl. Microbiol. Biotechnol.* **2012**, *96*, 885–94.
56. Hou, J. J. C.; Codamo, J.; Pilbrough, W.; Hughes, B.; Gray, P. P.; Munro, T. P. New frontiers in cell line development: challenges for biosimilars. *J. Chem. Technol. Biotechnol.* **2011**, *86*, 895–904.
57. Shukla, A. a; Gottschalk, U. Single-use disposable technologies for biopharmaceutical manufacturing. *Trends Biotechnol.* **2013**, *31*, 147–54.
58. Read, E. K.; Park, J. T.; Shah, R. B.; Riley, B. S.; Brorson, K. a; Rathore, a S. Process analytical technology (PAT) for biopharmaceutical products: Part I. Concepts and applications. *Biotechnol. Bioeng.* **2010**, *105*, 276–84.
59. Kelley, B.; Blank, G.; Lee, A. Downstream Processing of Monoclonal Antibodies: Current Practices and Future Opportunities. In *Process Scale Purification of Antibodies*; Gottschalk, U., Ed.; Wiley: Hoboken, NJ, 2009; pp 1–24.
60. Shukla, A. A.; Hubbard, B.; Tressel, T.; Guhan, S.; Low, D. Downstream processing of monoclonal antibodies: Application of platform approaches. *J. Chromatogr. B* **2007**, *848*, 28–39.
61. Fahrner, R.; Knudsen, H.; Basey, C.; Galan, W.; Feuerhelm, D.; Vanderlaan, M.; Blank, G. Industrial purification of pharmaceutical antibodies: Development, operation, and validation of chromatography processes. *Biotechnol. Genet. Eng. Rev.* **2001**, *18*, 301–327.
62. Ghose, S.; Hubbard, B.; Cramer, S. M. Binding capacity differences for antibodies and Fc-fusion proteins on protein A chromatographic materials. *Biotechnol. Bioeng.* **2007**, *96*, 768–779.
63. Tugcu, N.; Roush, D. J.; Göklen, K. E. Maximizing productivity of chromatography steps for purification of monoclonal antibodies. *Biotechnol. Bioeng.* **2008**, *99*, 599–613.
64. Ghose, S.; Al, E. USe and optimization of dual flow-rate loading strategy for maximizing throughput in protein A affinity chromatgrpahy. *Biotechnol. Prog.* **2004**, *20*, 830–840.
65. Jiang, C.; Liu, J.; Rubacha, M.; Shukla, A. A. A mechanistic study of Protein A chromatography resin lifetime. *J. Chromatogr. A* **2009**, *1216*, 5849–5855.
66. Shukla, A. A.; Hinckley, P. Host cell protein clearance during Protein A chromatography: Development of an improved column wash step. *Biotechnol. Prog.* **2008**, *24*, 1115–1121.
67. Levy, N. E.; Valente, K. N.; Choe, L. H.; Lee, K. H.; Lenhoff, A. M. Identification and characterization of host cell protein product-associated impurities in monoclonal antibody bioprocessing. *Biotechnol. Bioeng.* **2013**, *9999*, 1–9.
68. Ghose, S.; Jin, M.; Liu, J.; Hickey, J. Integrated Polishing Steps for Monoclonal Antibody Purification. In *Process Scale Purification of Antibodies*; Gottschalk, U., Ed.; John Wiley and Sons: Hoboken, NJ, 2009; pp 145–168.

69. Tscheliessnig, A. L.; Konrath, J.; Bates, R.; Jungbauer, A. Host cell protein analysis in therapeutic protein bioprocessing: Methods and applications. *Biotechnol. J.* **2013**, *8*, 655–70.
70. Barroso, T.; Hussain, A.; Roque, A. C. a; Aguiar-Ricardo, A. Functional monolithic platforms: Chromatographic tools for antibody purification. *Biotechnol. J.* **2013**, *8*, 671–81.
71. Kallberg, K.; Johansson, H.-O.; Bulow, L. Multimodal chromatography: An efficient tool in downstream processing of proteins. *Biotechnol. J.* **2012**, *7*, 1485–95.
72. Wang, X.; Hunter, A. K.; Mozier, N. M. Host cell proteins in biologics development: Identification, quantitation and risk assessment. *Biotechnol. Bioeng.* **2009**, *103*, 446–58.
73. Schenauer, M. R.; Flynn, G. C.; Goetze, A. M. Identification and quantification of host cell protein impurities in biotherapeutics using mass spectrometry. *Anal. Biochem.* **2012**, *428*, 150–7.
74. Valente, K. N.; Schaefer, A. K.; Kempton, H. R.; Lenhoff, A. M.; Lee, K. H. Recovery of Chinese hamster ovary host cell proteins for proteomic analysis. *Biotechnol. J.* **2014**, *9*, 87–99.
75. Jeong, S. H. Analytical methods and formulation factors to enhance protein stability in solution. *Arch. Pharm. Res.* **2012**, *35*, 1871–86.
76. *Q8(R2). Pharmaceutical Development.* International Conference on Harmonisation of Technical Requirements for Registration of Pharmaceuticals for Human Use (ICH), 2009.
77. *A-Mab: A Case Study in Bioprocess Development*, version 2.1; CMC Biotech Working Group, 2009.
78. Justice, C.; Brix, A.; Freimark, D.; Kraume, M.; Pfromm, P.; Eichenmueller, B.; Czermak, P. Process control in cell culture technology using dielectric spectroscopy. *Biotechnol. Adv.* **2011**, *29*, 391–401.
79. Rathore, a S.; Bhambure, R.; Ghare, V. Process analytical technology (PAT) for biopharmaceutical products. *Anal. Bioanal. Chem.* **2010**, *398*, 137–54.
80. Streefland, M.; Martens, D. E.; Beuvery, E. C.; Wijffels, R. H. Process analytical technology (PAT) tools for the cultivation step in biopharmaceutical production. *Eng. Life Sci.* **2013**, *13*, 212–223.
81. Awotwe-Otoo, D.; Agarabi, C.; Khan, M. a. An integrated process analytical technology (PAT) approach to monitoring the effect of supercooling on lyophilization product and process parameters of model monoclonal antibody formulations. *J. Pharm. Sci.* **2014**, *103*, 2042–52.
82. Mercier, S. M.; Diepenbroek, B.; Wijffels, R. H.; Streefland, M. Multivariate PAT solutions for biopharmaceutical cultivation: urrent progress and limitations. *Trends Biotechnol.* **2014**, *32*, 329–336.
83. Read, E. K.; Shah, R. B.; Riley, B. S.; Park, J. T.; Brorson, K. a; Rathore, a S. Process analytical technology (PAT) for biopharmaceutical products: Part II. Concepts and applications. *Biotechnol. Bioeng.* **2010**, *105*, 285–95.
84. Federici, M.; Lubiniecki, A.; Manikwar, P.; Volkin, D. B. Analytical lessons learned from selected therapeutic protein drug comparability studies. *Biologicals* **2013**, *41*, 131–47.

85. Douette, P.; Bolon, P. Analytical Method Lifecycle : A Roadmap for Biopharmaceutical development. *Biopharm Int.* **2013**, 46–53.
86. *Q2(R1) Validation of Analytical Procedures*. International Conference on Harmonisation of Technical Requirements for Registration of Pharmaceuticals for Human Use (ICH), 2005.
87. Blood Products and Related Biologicals. World Health Organization. http://www.who.int/bloodproducts/ref_materials/en/ (accessed March 7, 2014).
88. Williams, R. L. Official USP Reference Standards: Metrology concepts, overview, and scientific issues and opportunities. *J. Pharm. Biomed. Anal.* **2006**, *40*, 3–15.
89. Bhattacharyya, L.; Cecil, T.; Dabbah, R.; Roll, D.; Schuber, S.; Sheinin, E. B.; Williams, R. L. The value of USP public standards for therapeutic products. *Pharm. Res.* **2004**, *21*, 1725–31.
90. *USP Guideline for Submitting Requests for Revision to USP-NF*. U.S. Pharmacopeia Convention, 2011, Vol. 5, pp 1−5; http://www.usp.org/usp-nf/development-process/submit-new-monographs/submission-guidelines.
91. Rituximab Final Authorized, version 1.0. U.S. Pharmacopeia Convention. https://mc.usp.org/monographs/rituximab-1-0.
92. USP. <129> Analytical Procedures for Recombinant Therapeutic Monoclonal Antibodes. *Pharmacopeal Forum* **2013**, *39*, 2368–2368.
93. Rituximab. Monographs on Biological Products for Incorporation in IP-2014. National Institute of Biologicals. http://nib.gov.in/Monographs_developed.html (accessed March 7, 2014).
94. *Draft Guidance for Industry Bioanalytical Method Validation*. Federal Register, September 13, 2013.
95. Apostol, I.; Krull, I.; Kelner, D. Analytical Method Validation for Biopharmaceuticals. In *Analytical Chemistry*; InTech: Rijeka, Croatia, 2012; pp 115–134.
96. Rozet, E.; Marini, R. D.; Ziemons, E.; Boulanger, B.; Hubert, P. Advances in validation, risk and uncertainty assessment of bioanalytical methods. *J. Pharm. Biomed. Anal.* **2011**, *55*, 848–58.
97. Schiel, J. E.; Au, J.; He, H.-J.; Phinney, K. W. LC-MS/MS biopharmaceutical glycoanalysis: Identification of desirable reference material characteristics. *Anal. Bioanal. Chem.* **2012**, *403*, 2279–89.
98. Terms and Definitions Used in Connection with Reference Materials. *ISO Guide 30:1992/Amd 1:2008*; ISO (International Organization for Standardization): Geneva, Switzerland, 2008.
99. Reichert, J. M. Antibodies to watch in 2013: Mid-year update. *mAbs* **2013**, *5*, 513–7.

Chapter 2

Monoclonal Antibodies: Mechanisms of Action

Roy Jefferis*

School of Immunity and Infection,
College of Medical and Dental Sciences, University of Birmingham,
Edgbaston, Birmingham B15 2TT, United Kingdom
*E-mail: r.jefferis@bham.ac.uk

We live in a hostile environment and are dependent for protection on the innate and adaptive immune systems. A major component of these systems is antibody molecules, which bind pathogens with exquisite specificity to form immune complexes that activate downstream mechanisms, leading to pathogens' removal and destruction. Five classes (nine isotypes) of human antibody have been identified. The immunoglobulin G (IgG) class predominates in serum and a majority of monoclonal antibody (mAb) therapeutics are based on the IgG format. Selection within the antibody repertoire allows the generation of mAbs having specificity for any selected target, including human (self) antigens, and genetic engineering allows the development of any chosen isotype. This review focuses on the structure and function of the four human IgG isotypes (subclasses) and the biologic functions that their immune complexes activate through interactions with cellular Fc receptors (FcγR and FcRn) and/or the C1q component of complement. The long catabolic half-life (~21 days) of IgG contributes to its efficacy as a therapeutic. The human IgG subclasses exhibit high sequence homology, but each exhibits a unique profile of biologic activities that are modulated with the glycoform profile of the IgG-Fc. A comprehensive appreciation of the structure−function relationships for native serum derived IgG allows for protein and glycosylation engineering to enhance or eliminate biologic activities and the potential for the generation of mAb therapeutics that are optimal for a given disease indication.

© 2014 American Chemical Society

Introduction

The commonly perceived "hallmark" of an antibody is its specificity for a target pathogen (antigen); however, protection is dependent also on the activation of a cascade of downstream biologic mechanisms, triggered by the antigen–antibody (immune) complexes formed, resulting in the killing and elimination of pathogenic organisms or pathologic targets. The seminal publication of Kohler and Milstein in 1975 (*1*) reported the establishment of the hybridoma technique and the generation of monoclonal antibodies (mAbs) of predetermined antigen specificity. Initially, these antibodies were of mouse origin and not suitable for *in vivo* applications in man. Subsequent developments in genetics and protein engineering provided tools for the generation of chimeric mouse/human, humanised and fully human antibodies that have received regulatory authority approval as therapeutics. The antibody format is constantly being expanded to develop therapeutics designed to be optimal for given disease indications. Structural and functional characterisation of this diverse array of therapeutics is a challenge that is being addressed by the contributors to this volume. Whilst the protocols discussed in succeeding chapters are of general utility in characterising protein therapeutics, the focus is characterisation of a full-length recombinant IgG antibody mAb, comprised of a heavy chain of gamma subclass one ($\gamma 1$) and a light chain of type kappa (κ). It is intended to fully characterise this mAb, employing multiple orthogonal analytical techniques, with a view to establishing it as a Standard Reference Material (*2*).

The development and clinical application of "Small Molecule Drugs" (Mass < 700 Da) has a long and successful history. During the period in which an innovator company has patent protection, competitor companies can develop an identical drug product that may subsequently be approved and marketed as a generic pharmaceutical. Pharmacovigilance exercised throughout the life of a drug ensures that clinical efficacy is maintained and that any post-approval adverse events are reported. Confidence in this process has been achieved through decades of collaboration between pharmaceutical companies and regulatory authorities. By contrast, the subject of these volumes is recombinant antibody therapeutics that are complex biologics that cannot be synthesised chemically and are necessarily produced in living cells, the drug product being harvested and subjected to extensive downstream purification and formulation protocols (*3–5*). In addition to the demonstration of clinical efficacy, an innovator company seeking approval for a potential antibody therapeutic is required to characterise the drug product—biochemically and biophysically—using multiple orthogonal technologies. The parameters established define the drug product and must be maintained throughout the lifetime of the drug. In the process, Critical Quality Attributes (CQAs) are identified that ensure drug efficacy and can be achieved employing Quality by Design (QbD) parameters unique to the production platform and downstream protocols employed. The parameters defining CQAs and QbD are the undisclosed intellectual property of the innovator company; consequently, it is deemed essentially impossible to produce an identical product within another facility (i.e., in principle, it is not possible to develop generic biopharmaceuticals) (*3–5*). It is often claimed that "The process defines the product".

Initially, innovator companies generating mAb therapeutics emphasized their large size, structural complexity and unavoidable heterogeneity and, hence, the impossibility of another provider being able to generate an identical "copy" of innovator product. However, as mAbs achieved "blockbuster" status and the period of patent protection shortened, "Big Pharma" recognised the financial potential of a share in these "blockbuster" markets. Consequently, pharmaceutical companies, large and small, have programmes to generate "copies" of successful antibody therapeutics. It is recognised that these "copies" will not be identical to innovator product, but regulatory authorities require that they be demonstrated to be "comparable" (*6*, *7*). If approved, they are not designated as generics but as "biosimilars" (Europe) or "follow-on biologics" (United States). The criteria for approval of a biosimilar biologic differ among national regulatory authorities, and biosimilar antibody drugs that have been approved in India (Reditux/Rituxan) and South Korea (Remsima/Remicade) have not been automatically approved by the European Medicines Authority (EMA) or the U.S. Federal Drug Administration (FDA). It had been rumoured that biosimilar programmes have progressed slowly due, in part, to a perceived uncertainty and lack of confidence in regulatory approval pathways. However, Remsima (Remsima, Celltrion; Inflectra, Hospira), a biosimilar candidate for Remicade developed under EMA guidelines was approved by the EMA in 2013 (*8*, *9*). Approval was heralded as a "landmark" event and demonstrates that the EMA has confidence in its ability to evaluate the comparability of biosimilar antibody products; this may provide incentive to the biosimilar "industry" and for the FDA to finalise and publish definitive guidelines.

Human Antibody Isotypes

An intact immune system is essential to the integrity of the individual. We live in a hostile environment and are constantly exposed to potential pathogens; however, we are mostly unaware of these insults due to protective innate and adaptive immune responses (*10–12*). The adaptive immune system has many components, but the focus in this review is the role of antibodies. Initial contact with antigen provokes a primary immune response and the production of antibodies of the immunoglobulin M (IgM) class. Further stimulation by antigen may lead to a secondary response, characterized by the production of antibody of the IgG, IgA, IgE, and/or IgD class, a process referred to as class or isotype switching. There are four subclasses of IgG (IgG1, IgG2, IgG3, and IgG4) and two subclasses of IgA (IgA1 and IgA2), giving a total of nine antibody isotypes in humans (*12–15*). Each antibody isotype exhibits unique structural and functional properties. In addition, Ig genes are polymorphic, and there are quantitative and qualitative differences in haplotype (allotype) distribution between population groups (*16*). Secondary immune responses also establish memory that provides for a rapid and amplified response to subsequent contact with the same pathogen. Protective antibody responses to environmental antigens are polyclonal and, consequently, heterogeneous due to the production of antibodies specific for each of the many epitopes expressed on the antigen, and may be comprised of multiple isotypes. It is not possible, therefore, to unequivocally define the mechanism(s) of action that results in immune protection (e.g., neutralization, lysis, phagocytosis). However, IgG is quantitatively the predominant antibody isotype present in normal human serum and is the isotype most studied to elucidate the relationship between structure and function. To date, all licensed intact mAb therapeutics have been based on the IgG format and produced by transfected Chinese hamster ovary (CHO), mouse NSO, or mouse Sp2/0 cells. The quality of the antibody therapeutic product will depend on the ability of the chosen cell type, and individual cell line, to effect post-translational modifications similar or identical to those of human plasma cells. Initially, IgG subclass selection was informed by the accumulated understanding of natural antibody responses; however, research within the pharmaceutical industry and clinical experience is approaching a maturity that allows for definitive assignment of effector functions to a given IgG mAb, dependent on the subclass, epitope specificity, glycoform, and structure of immune complexes formed (*5, 12–14, 17, 18*).

The Polypeptide Structure of Human IgG

The characteristic four chain homodimer structure of IgG antibodies was established in the 1950s and the contributions of Rodney Porter (UK) and Gerald Edelman (United States) recognised with the award of the Nobel Prize in 1972. The Edelman laboratory published the complete covalent structure of a monoclonal human IgG1 subclass protein (Eu, IgG1κ), isolated from the serum of a patient having multiple myeloma (*19*). This protein defined the sequence and numbering of amino acid residues of both the heavy and light chains (e.g., asparagine 297 (N297) as the attachment site for oligosaccharide). These residue assignments are perpetuated although the length of the heavy and light chains varies among antibodies. At the protein sequence level, the light (25kDa) and heavy (50kDa) chains are comprised of two and four sequence homology regions of ~110 amino acid residues, respectively (see Figure 1); at the gene level, each homology region is encoded within an exon separated by intervening introns. Each homology region folds to form a β-barrel structure comprised of two anti-parallel β-pleated sheets bridged by an intrachain disulphide bond, connected through β-bends; hydrophobic side chains are orientated toward the interior whilst hydrophilic side chains are more exposed to solvent (*20, 21*). This stable protein "scaffold" is referred to as the Ig fold; it is widely used within the proteome and allows for virtually unlimited sequence variation, particularly within the β-bends, and the generation of unique interaction or receptor sites (*12–15*).

The N-terminal homology regions of the light (V_L) and heavy (V_H) chains can differ in length between antibodies, and exhibit unique sequences that determine the epitope specificity of the intact antibody. Maximum sequence diversity is localised within three hypervariable or complementarity-determining regions (CDRs), situated at β-bends, of both the heavy and light chains that are brought into spatial proximity by the Ig fold to form a unique antigen (epitope) binding site (paratope) (*15*). Humans express two isotypes of light chain, kappa and lambda (λ), and four gamma IgG heavy chain isotypes or subclasses (γ1, γ2, γ3, and γ4), encoded by genes on chromosomes 2, 22, and 14, respectively. Each light chain is characterised by one constant homology domain, C_L (C_κ or C_λ), and each heavy chain by three constant homology regions, C_H1, C_H2, and C_H3. The C_κ and C_λ domains each bind with the heavy chain C_H1 domain through multiple noncovalent interactions and a single interchain disulphide bridge. Plasma cells express only one heavy chain and one light chain gene to secrete homodimer antibodies that are comprised of either $[V_H/V_L\text{-}C_\kappa/C_H1\text{-}h\text{-}C_H2\text{-}C_H3]2$ or $[V_H/V_L\text{-}C_\lambda/C_H1\text{-}h\text{-}C_H2\text{-}C_H3]2$ (h; hinge) homology regions. Formation of the homodimer is dependent of the formation of a single disulphide bridge between the heavy and light chains, multiple inter-heavy chain disulphide bridges within the hinge region, multiple *trans* noncovalent interactions between the C_H3 domains, and lateral noncovalent interactions at the C_H2/C_H3 interface (*12–15, 19–21*) (see Figure 2). The formation of intra- and inter-disulphide bonds has been shown to be a source of structural heterogeneity both for serum-derived IgG and recombinant mAbs, particularly the IgG2 subclass (*22*). The hinge region allows for independent mobility of the fragment antigen binding (Fab) and fragment crystallisable (Fc) moieties such that the intact IgG molecule is functionally divalent for antigen binding, with the formation of antigen–antibody (immune) complexes, whilst the Fc is accessible to engage ligands that initiate effector pathways (*12–15, 17–22*).

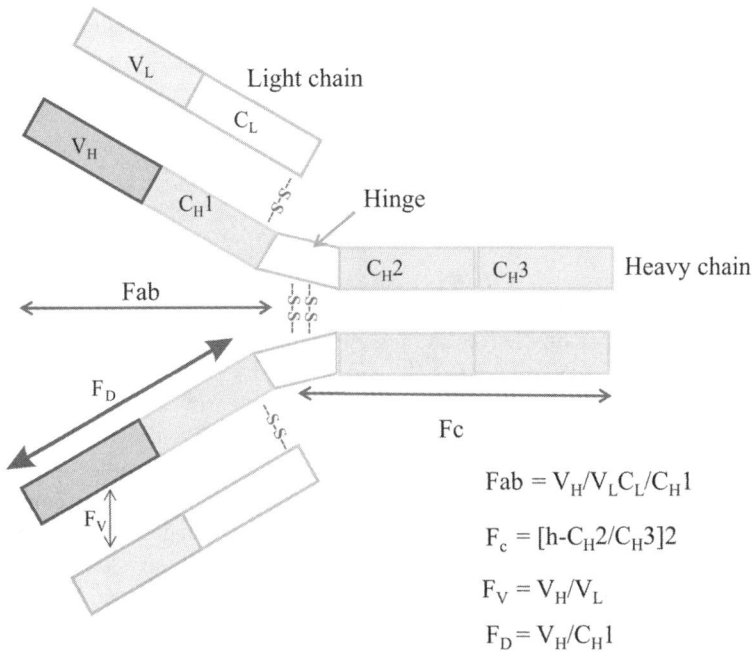

Figure 1. A cartoon of the four chain structure for an IgG1 molecule. Only interchain disulphide bridges are shown. Subscripts H and L refer to antibody heavy and light chains; V and C refer to variable and constant domains; and Fab, Fv, and Fc represent antigen-binding, variable, and crystallisable fragments.
(see color insert)

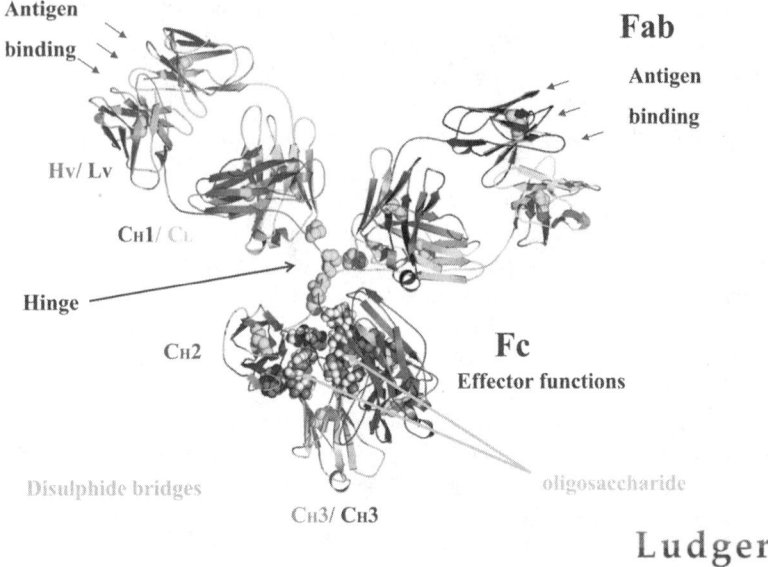

Figure 2. The domain structure of the IgG molecule. Fab and Fc refer to antigen-binding and crystallisable moieties. (see color insert)

The enzyme papain cleaves the heavy chain in the hinge region to release V_H/V_L-C_κ/C_H1 or V_H/V_L-C_λ/C_H1 (Fab) and $[h$-$C_H2/C_H3]2$ (Fc) protein fragments (Figure 1). The Fab fragment retains structural integrity and antigen-binding activity due to domain pairing, with the formation of multiple noncovalent interactions, between V_H/V_L and C_H1/C_L domains. In the intact molecule, the two Fab structures are linked to the Fc through a linker or "hinge" region comprised of flexible "upper" and "lower" regions. The upper and lower regions are linked through a rigid core comprised of proline and cysteine residues, the latter forming inter-heavy chain disulphide bridges. The Fc also forms an independent protein moiety stabilised by noncovalent C_H3/C_H3 pairing, lateral C_H2/C_H3 interactions, and inter-heavy chain disulphide bridges in the hinge region. The C_H2 domains do not pair but form an open horseshoe structure, with much of the internal space occupied by a complex diantennary oligosaccharide structure. The oligosaccharide is attached through a covalent bond to asparagine 297 (Eu) and, in addition, forms multiple noncovalent interactions with the polypeptide backbone and amino acid residue side chains of the inner surface of the C_H2 domain (shown in Figure 2).

Human IgG Gene Polymorphism: Allotypes and Idiotypes

Human IgG polymorphism (allotypes) was first described by Grubb who discovered that certain human sera would agglutinate erythrocytes sensitized with human incomplete anti-Rh antibody (*16, 23, 24*). Extensive (~5 %) polymorphism (allotypy) within human IgG heavy chains and kappa light chains was subsequently recognised by careful serological typing using human reagents obtained from multiparous women, multiple transfused individuals, or normal blood donors. Thus, the discovery of this polymorphism demonstrates that exposure of an individual to IgG of a non-self allotype can induce an anti-allotype response. Gene sequencing studies have revealed further extensive structural polymorphisms that do not appear to provoke an immune response (*24*).

By definition, allotypes are shared among groups of individuals within an outbred population; however, studies in rabbits showed that antibodies raised within an individual rabbit could be immunogenic when administered as an antigen−antibody immune complex to a recipient rabbit of the same allotype. The unique epitopes recognised were termed "o" and subsequently shown to be serological markers for the unique structure and specificity of an antibody, particularly, but not exclusively, the antigen-binding site (paratope). Historically, further research of the phenomenon of idiotypy led to the proposal of immune regulation through an idiotype network, with subsequent esoteric arguments and controversy (*25*). The term idiotype is now in common use to define the serological uniqueness of a mAb. The idiotype of a mAb may be the target for an anti-idiotype response in patients, resulting in neutralisation of its antigen-binding activity with loss of efficacy and possible further adverse events.

In 1976 the World Health Organization sponsored an expert committee at which the nomenclature for human Ig allotypes was systematized and a numerical system was proposed to replace the original alphabetic system (*26, 27*). Both systems may be encountered in the literature, particularly when reference is made to original publications in which the allotypes were described (see Table 1). Allotypes of IgG proteins were originally defined by the expression of unique epitope(s) recognised by unique serologic reagent(s); however, it is now possible to assign allotype by gene or protein sequencing (*28–32*). Allotypes expressed on the constant region of IgG heavy chain are designated as Gm (genetic marker) together with the subclass (e.g., G1m) and the allotype number (or letter) (e.g., G1m(1) or G1m(a), G3m(5) or G3m(b1)). Polymorphisms within IgA and kappa light chains are designated A2m (e.g., A2m(1)) and Km (e.g., Km1), respectively. Serological polymorphisms have not been reported for lambda chains; however, there are multiple lambda chain isotypes and the number of IGLC (C_λ) genes can vary between individuals (*28, 29*). Because the multiple genes encoding the constant region of the heavy chains (IGHC) are closely linked within the IGH gene locus, they are inherited together as a haplotype with a low frequency of crossovers. Crossover events have occurred during evolution with the result that present population groups may express a characteristic haplotype (*16, 23, 24, 28–32*).

Table 1. Human IgG allotypes

	Heavy Chains				Light Chains
Isotype/type	IgG1	IgG2	IgG3	IgA	κ
Allotypes	G1m	G2m	G3m	A2m	Km
	1 (a) 2 (x)	23 (n)	21 (g1) 28 (g5)	1 2	1 2
	3 (f) 17 (z)		11 (b0) 5 (b1)		3
			13 (b3) 14 (b4)		
			10 (b5) 15 (s)		
			16 (t) 6 (c3)		
			24 (c5) 26 (u)		
			27 (v)		

Allotype Expression and Amino Acid Correlates (Eu Numbering)

IgG1

The heavy chains of IgG1 proteins may express G1m(3), G1m(17,1), or G1m(17,1,2) allotypes, respectively (*16, 28–32*). The amino acid residue correlates are G1m(1), D356/L358; non-G1m(1), E356/M358; G1m(2), G431; non-G1m(2), A431, G1m(3), R214; and G1m(17), K214 (see Figure 3). The non-IgG1m(1) sequences are referred to as isoallotypes; the residues defined are common to other IgG isotypes and are not immunogenic (i.e., all individuals are tolerant and humans do not develop serological reagents). The non-G1m(2) sequence is similarly an isoallotype. The proposed Reference Standard mAb that is the subject of the NIST-sponsored characterisation exercise is a humanized IgG1κ molecule of G1m(3).Km3 allotype (Figure 3, Table 1). It was produced in mammalian cell culture and is supplied in 1 ml aliquots at 10 mg/mL or 100 mg/mL; it is formulated in 12.5 mM L-histidine or 12.5 mM L-histidine HCl monohydrate (pH 6.0).

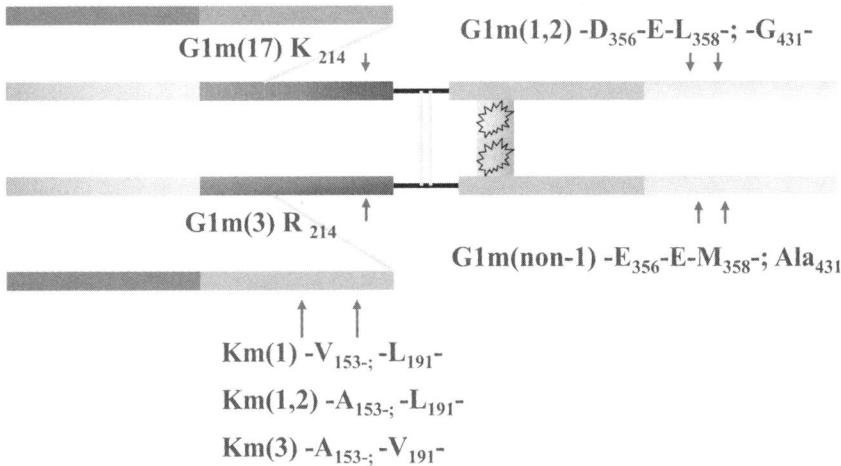

Figure 3. Sequence correlates of IgG1.G1m and Km allotypes. (see color insert)

IgG2

The heavy chains of IgG2 proteins may express the G2m(23) allotype, which is correlated with the presence of residues T189 and M282. The IgG2 allele encodes for P189 and V282; however, the latter residues are common to other IgG subclasses and represent isoallotypes. They are referred to as non-G2m(23) or G2m (*22, 31–33*).

IgG3

Multiple serologically defined IgG3 allotypes have been established; however, further complexity is evident at the DNA level (*29, 30*). Analysis of 19 DNA sequences of the G3m(5*) haplotype yielded 11 different allelic sequences; similarly analysis of 10 DNA sequences of the G3m(21*) haplotype yielded 4 distinct sequences (*30*). An additional polymorphism results from differing numbers of hinge region exons, varying from 2 to 5 exons, the first encoding a common hinge region sequence with 1 to 4 repeats of a second exon (*33, 34*). Thus, the genetic hinge region protein sequence can vary between 22 to 62 amino acid residues, with 3 to 11 inter-heavy chain disulphide bridges, that may significantly influence relative Fab versus Fc mobility and access to ligand binding sites (*31–34*).

IgG4

Serologically defined allotypes of IgG4 have not been reported; however, the nG4m(a) and nG4m(b) isoallotypic variants were defined structurally and correlate with L309 and V309, respectively. A further isoallotypic variant R/K409 has been shown to exhibit structural and functional significance, with the R409 variant being permissive of "Fab arm exchange" (*35*).

Kappa and Lambda Light Chains

In humans, the IgGκ to IgGλ expression ratio is approximately 60:40. The human genome has one kappa constant (IGKC) gene but variable number of lambda constant (IGLC) genes; therefore, there are two different forms of polymorphism, each of which may have consequences for immunogenicity of mAbs. There are three kappa chain allotypes—designated Km(1), Km(1,2), and Km(3)—which define three Km alleles. Expression of the Km1 epitope (allotope) correlates with residues V153/L191, Km(1,2) with A153/L191, and Km3 with A153/V191 (*28*, *31*, *33*) Table 1 and Figure 3). Serologically defined allotypes of the constant region of lambda chains have not been reported; however, the number of IGLC genes varies between 7 and 11, depending on individual haplotypes (*28–33*), and some of these isotypes can also be distinguished serologically.

Allotypy of Licensed Chimeric mAb Therapeutics

There are 6 amino acid residues differences between IgG1 proteins that are G1m(3).Km(1) and G1m(17,1,2).Km(3), respectively. This could constitute a significant antigenic "barrier" when a patient homozygous for one haplotype is exposed to a mAb of the alternate allotype. Biopharmaceutical companies producing and marketing mAbs rarely disclose the sequence of the constant regions of their mAbs, and hence the allotype, or give any rationale for their choice. Sequences may be available in data bases, but these have proved to be inaccurate, with different sources reporting different sequences (*16*). This reflects earlier research imperatives that concentrated on the development of genetic engineering techniques for the generation of chimeric and humanized antibodies, and companies obtained IGHC and IGLC genes from an available source. Currently, IgG1 mAb therapeutics of both predominant allotypes are licensed and in the market place (*16*).

A panel of licensed mAbs were serologically tested to determine the allotype distribution; both G1m(3) and G1m(17,1) heavy chain allotypes were shown to be employed (*16*). An apparently anomalous result was obtained for trastuzumab (Herceptin) and omalizumab (Xolair). These proteins appeared, serologically, to have heavy chains that were G1m(17) only, rather than the expected G1m(17,1) or G1m(17,1,2) haplotype. A literature search revealed that Genentech engineered an original G1m(1,17) gene to replace the D356/L358 G1m(1) sequence with the isoallotype nG1m(1) sequence E356/M358, with the objective of reducing the potential for immunogenicity (*36*). This approach has been extended to generate an IgG1 protein with a "null" allotype sequence by the additional replacement of C_H1 arginine 214 (G1m(3)) or lysine 214 (G1m(17)), respectively, with threonine (*37*).

The Quaternary Structure of Human IgG

Crystal structures for 10 intact IgG molecules have been reported (*12*); however, resolution of both Fab and Fc structures was obtained for only two (*21, 38, 39*). One was essentially a rheumatoid factor (RF)-like auto-antibody in which the Fab had specificity for a Fc epitope (*38, 39*); the other resulted from stabilisation of structure due to the close proximity (contact!) between the Fab and the Fc of a neighbouring molecule (*21*). Because crystallisation depends on the formation of protein−protein interactions and data is collected at temperatures of ~100 K. X-ray crystallography presents a rather static structural model. More dynamic insights may be gained employing techniques operated at ambient temperatures, for example, X-ray scattering (*40–42*) and nuclear magnetic resonance (NMR) (*43, 44*). The length of the hinge region and the number of inter-heavy chain disulphide bonds differs significantly between the human IgG subclasses, influencing mobility and average solution conformation of the IgG-Fab and IgG-Fc moieties with respect to each other. This may include the ability to assume a "dislocated" form that provides access for Fc receptors (FcγR) and the C1 component of complement-to-effector ligand-binding sites localised to the hinge-proximal region of the C_H2 domain (*12–15*). Engineering the extended hinge region of IgG3 molecules to generate hinge regions of different lengths revealed no direct relationship between hinge length and the ability to bind and activate the C1 component of complement; however, at least one inter-heavy chain disulphide bridge was shown to be essential (*45*).

Quaternary Structure of IgG-Fc: The Protein Moiety

The first crystal structure of IgG-Fc, resolved at 2.9 Å, was published by Deisenhofer in 1981 (*20*). The IgG-Fc was generated by papain cleavage of polyclonal IgG at the Lys 222−Thr 223 peptide bond, within the hinge region, and extending to a C-terminus residue at 446. It was reported that interpretable electron density was not obtained for residues 223 through 237, which comprise most of the core and lower hinge region, or the C-terminal residues 444 through 446; it was not known at that time that the C_H3 exon codes for a C-terminal 447 lysine residue that is removed by endogenous serum carboxypeptidase B. Similar structures have been reported for human, rabbit, and mouse IgG-Fc, as well as chicken IgY-Fc fragments, (*12–15*) and have provided further structural insight due to progressively higher resolution. An α-carbon IgG ribbon structure is shown in Figure 2 (courtesy of R. Emery and D. Fernandez, Ludger UK; based on the pdb 1IGY molecule) (*12*). The common structural features reported are:

- The C_H3 domains are well defined due to noncovalent pairing, involving ~2000 Å² of accessible surface area in the $(C_H3)2$ module.
- The area of noncovalent contact between the C_H2 and C_H3 domains is ~800 Å². This suggests that the C_H2-C_H3 contact contributes to the relative stability observed for the C-terminal proximal region of C_H2 domains, as opposed to the "softness" of the C_H2 domain proximal to the hinge region.

- The C_H2 domains do not pair and the hydrophobic surface of each C_H2 domain is "overlaid" by the carbohydrate. Hydrophobic and polar interactions between the oligosaccharide and the C_H2 domain surface occupy ~500 Å^2 and substitute for domain pairing (20, 46).
- One C_H2 domain was more ordered than the other, due to its crystal contact with a neighbouring C_H2 domain.
- A more disordered structure for the hinge-proximal region of the C_H2 domain is reflected in higher temperature factors.
- The intrinsic stability of the Ig fold is reflected in higher structural resolution for the β-sheets regions than for than for β-bends.

These interpretations and conclusions have been confirmed in x-ray structures obtained for human IgG-Fc alone or in complex with Staphylococcal protein A (SpA); Streptococcal protein G (SpG); RF; and human sFcγRIIa, sFcγRIIIb, and sFcγRIIIa (12–15). The observed internal mobility of the lower hinge and hinge-proximal regions of the C_H2 domains allows for an equilibrium of high-order conformers to be formed that may differentially bind unique ligands, for example, the three homologous Fcγ receptor types. Previous proposals that different ligands may bind through "overlapping non-identical sites" may suggest too rigid a structure (12, 47, 48) and may be modified to suggest that each FcγR binds to a unique IgG-Fc conformer present within an equilibrium of transient protein structures; however, amino acid residue side chains and/or main chain atoms may be involved in common (47, 48). The binding sites for sFcγRIIa, sFcγRIIIa, and sFcγRIIIb are asymmetric, with both heavy chains being engaged such that monomeric IgG is univalent for Fcγ receptors and the C1 component of complement; this obviates continuous activation of inflammatory cascades by circulating endogenous monomeric IgG *in vivo*; IgG antigen–antibody immune complexes are multivalent and able to cross-link and activate cellular receptors. Residues of the lower hinge region that are disordered in the Fc crystals are ordered in the Fc-FcγR complexes and directly involved in binding the receptor (49, 50). By contrast, IgG-Fc is functionally divalent for ligands binding at the C_H2-C_H3 interface, for example, the neonatal Fc receptor (FcRn), RF, SpA, and SpG. Due to the symmetry of the IgG-Fc, the two interaction sites are opposed at approximately 180°, and each is accessible to bind macromolecular ligands to form multimeric complexes.

The IgG-Fc Oligosaccharide Moiety

The IgG of human, rabbit, mouse, and other mammals have a consensus glycosylation sequon within the IgG-Fc, at N297 (Eu sequence) (19), and it has been demonstrated that the presence of a core diantennary heptasaccharide at this asparagine residue is essential for optimal activation of Fcγ receptors and the C1 component of complement. The oligosaccharide of normal polyclonal IgG-Fc is heterogeneous and essentially comprised of a core heptasaccharide with variable addition of fucose, galactose, bisecting N-acetylglucosamine, and sialic acid residues (see Figure 4) (51–55); recent analyses employing high sensitivity mass spectrometry suggests the possible presence of further minor oligosaccharides, for example, high mannose and hybrid glycoforms (54).

Several systems of nomenclature are currently in use to represent oligosaccharide structures; carbohydrate chemists, glycobiologists, and mass spectrometry scientists, among others, have developed different systems of nomenclatures (56, 57). Amongst antibody "practitioners", a shorthand nomenclature has evolved for oligosaccharides released from normal polyclonal IgG. A core heptasaccharide, highlighted in blue in Figure 4, is designated G0 (zero galactose); a core bearing one or two galactose residues is designated G1 or G2, respectively. The core + fucose is designated G0F; the core + fucose + galactose is designated G1F, G2F, and so forth. When bisecting N-acetylglucosamine is present, a B is added (e.g., G0B, G0BF, G1BF); sialylation at the galactose residues is designated as G1FS, G2FBS, and so forth. The approximate composition of neutral oligosaccharides released from normal polyclonal human IgG-Fc is G0, 3%; G1, 3%; G2, 6%; G0F, 23%; G1F, 30%; G2F, 24%; G0BF, 3%; G1BF, 4%; and G2BF, 7% (49–53). It is important to define the glycoform of the intact IgG molecule (e.g., [G0/G1F], [G1F/G2BF]) because it has been shown that individual IgG molecules may be comprised of symmetrical or asymmetrical heavy chain glycoform pairs (55, 58); consequently, enhanced FcγRIIIa-mediated antibody-dependent cellular cytotoxicity (ADCC) may be observed for IgG in which one heavy chain only bears oligosaccharide devoid of fucose (58, 59); thus the [G0/G0F] glycoform could be as potent in ADCC as the [G0/G0] glycoform.

Minor oligosaccharide structures present in polyclonal IgG-Fc may be significant because each could be the predominant form present of an individual antibody secreted from a single plasma cell; analysis of monoclonal human IgG, isolated from the sera of patient having plasma cell cancer (multiple myeloma), has shown that the IgG-Fc glycoform profile of the paraprotein is essentially unique for each protein analysed (60–62). Subtle differences in oligosaccharide processing was also observed, with a preference for addition of galactose to the α(1-6) arm of IgG1-Fc and the α(1-3) arm of IgG2-Fc; for the IgG3 subclass, the arm preference correlated with allotype (60–62). These data suggest a critical balance between the conformation of the IgG-Fc and the steric requirements of glycosyltranferases that may be sensitive to niche environments within the Golgi apparatus.

The glycoform profile of total polyclonal IgG can vary significantly in health and disease, particularly in autoimmune and inflammatory diseases (12, 63–66). Methods have been developed that allow the glycoform profile of antigen-specific polyclonal IgG autoantibodies to be analysed, and significant differences in the glycoform profiles of IgG autoantibodies and the bulk IgG have been reported (67, 68). The oligosaccharide profiles of recombinant IgG proteins produced in mammalian cells are significantly influenced by the cell type, the culture method, and precise conditions employed; however, the [G0F/G0F] glycoform predominates (see below). Under conditions of stress (e.g., nutrient depletion, acid pH), deviant glycosylation may be observed (e.g., the presence of high-mannose forms and/or incomplete occupancy) (69).

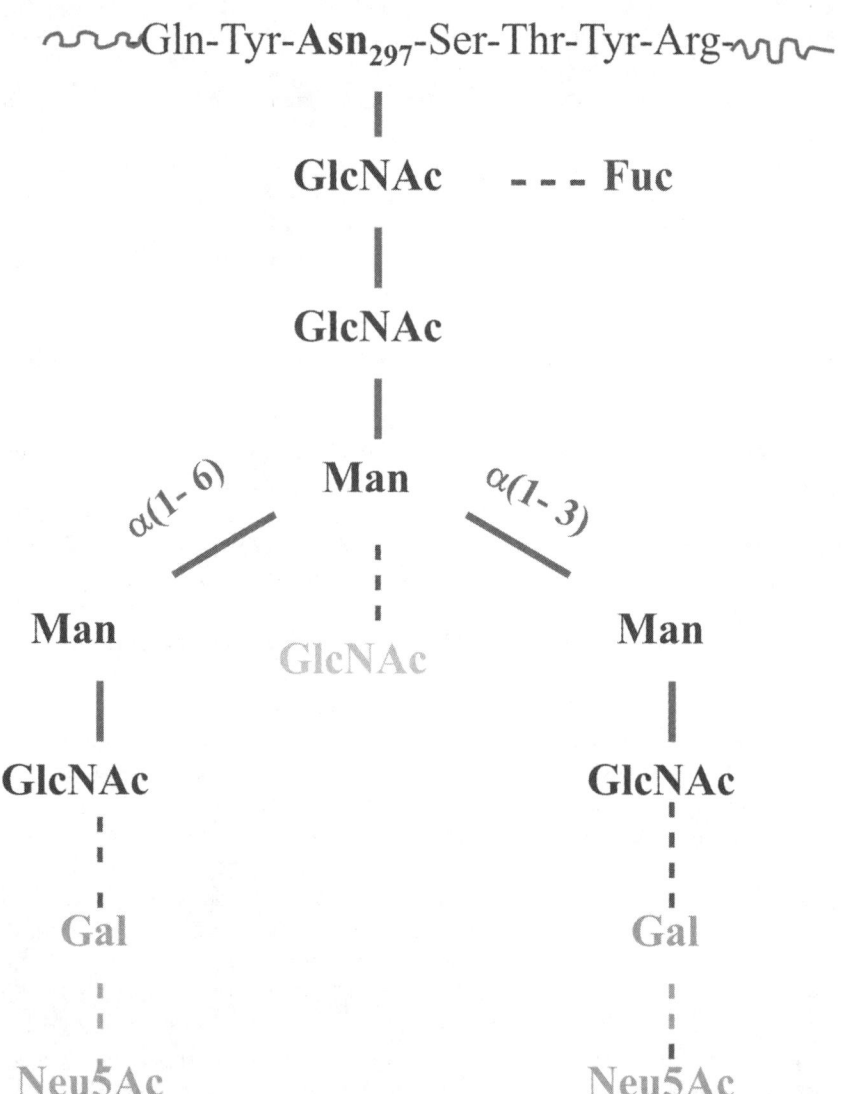

Figure 4. Representative IgG complex diantennary oligosaccharides comprised of a "core" -GlcNAc)2Man3(GlcNAc)2 heptasaccharide. (see color insert)

IgG-Fc Protein/Oligosaccharide Interactions

The Deisenhofer IgG-Fc structure indicated a potential for 72 protein−oligosaccharide interactions, including six C_H2 protein−oligosaccharide hydrogen bonds and six hydrogen bonds within each oligosaccharide moiety (*12, 20, 42–46*) (Figure 5). These interaction include the sugar residues of the α(1-6) arm, whilst residues of the α(1-3)-Man-GlcNAc arms are orientated toward the internal space between the C_H2 domains; weak lateral interactions between sugar residues present on opposed heavy chains have been suggested for some structures (20, 42–46, 70).

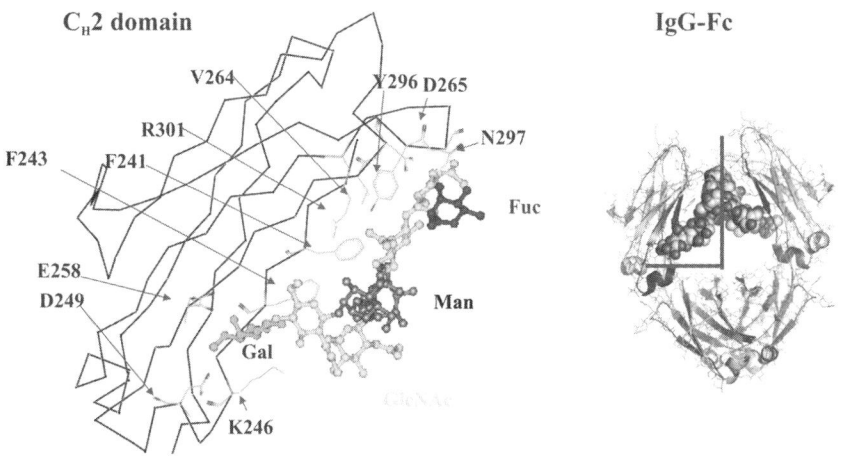

Figure 5. The IgG C_H2 domain, showing amino acid residues contributing noncovalent interactions with the oligosaccharide. (see color insert)

The properties of a series of normal, truncated, and aglycosylated glycoforms of IgG1-Fc, generated *in vitro*, were subjected to x-ray crystallographic analysis, differential scanning micro-calorimetry (DSMC) and Fcγ receptor binding (isothermal micro-calorimetry) (71–73). DSMC of [G2]2 and [G0]2 glycoforms exhibited two transition temperatures, Tm1 and Tm2, of 71.4°C and 82.2°C, representing the unfolding of the C_H2 and C_H3 domains, respectively. These data suggest that whilst the galactose residue on the α(1–6) arm has substantial contacts with the protein structure, it does not impact C_H2 domain stability. Sequential removal of the terminal N-acetylglucosamine and the two arm mannose residues, generating a [GlcNAc2Man]2 glycoform, resulted in destabilisation of the C_H2 domain and a lowering of Tm1 to 67.3°C, Tm2 remaining unchanged. The thermodynamic parameters describing C_H2 thermal denaturation of all IgG-Fc glycoforms was consistent with a cooperative unfolding, whilst the unfolding of the C_H2 domain of aglycosylated IgG1-Fc was non-cooperative, involving at least one intermediate (71–73). It was proposed that this intermediate is a partially unfolded C_H2 domain pair possessing hinge-proximal disordered or unfolded loops that may account for the compromised functional activities of deglycosylated IgG and IgG-Fc. DSMC has proved to be a valuable tool for probing the contributions of buffers and Fab sequence to stability and solubility of intact IgG and antibody fragments (74–76)

X-ray crystallography of the truncated IgG-Fc glycoforms revealed a progressive increase in temperature factors for the protein moiety of the C_H2 domain, evidence of increasing structural disorder (destabilisation) (71–73). Minimal (weak) FcγRI and C1 binding and activation was observed for the

[GlcNAc2Man]2 glycoform, which has the potential to form 31 noncovalent contacts with the protein, including at least three hydrogen bonds (*46*). Truncation of the sugar residues results in the mutual approach of C_H2 domains, with the generation of a "closed" conformation, in contrast to the "open" conformation observed for the fully galactosylated IgG-Fc (*73*). The dramatic reduction of FcγR and C1 binding and activation for aglycosylated IgG-Fc contrasts with consistent reports of minimal structural change within the protein structure.

An extensive NMR study of a series of truncated glycoforms showed that trimming of the oligosaccharide was accompanied by concomitant increase in the number of amino acid residues perturbed within the C_H2 domains (*43*). Cleavage between the primary and secondary N-acetylglucosamine sugar residues induced conformational changes within the lower hinge region, at sites that have no direct contact with the carbohydrate moieties, but form the major FcγR-binding site. Conformation at the C_H2/C_H3 interface, which forms the FcRn and protein A binding sites, was minimally perturbed. A dynamic model was also proposed from an NMR study of differentially galactosylated and sialylated IgG-Fc glycoforms (*44*). It was proposed that interactions of sugar residues of the α(1-6) arm with the protein surface may be a dynamic equilibrium between the bound and unbound state; the latter state may allow for increased accessibility to glycosyltransferases.

Although the oligosaccharide is integral to the IgG-Fc structure and appears to be sequestered within the space between the C_H2 domains microorganisms produce endoglycosidases that cleave the oligosaccharides from native IgG-Fc. Thus, *Streptococcus pyogenes* produces endoglycosidase S (EndoS), which cleaves between the primary and secondary N-acetylglucosamine residues (*77, 78*) Peptide N-Glycosidase F (PNGase F), isolated from culture filtrate of *Flavobacterium meningoscepticum*, cleaves the peptide/oligosaccharide bond to generate deglycosylated IgG (*54, 79*). Because the consequence of oligosaccharide cleavage is loss of effector functions and an ability to kill and remove target bacteria, it is tempting to conclude that this is evidence of co-evolution.

IgG-Fc Glycoform Profiles of Recombinant IgG Antibody Therapeutics

Glycosylation of IgG-Fc has a profound influence on the range and magnitude of effector functions activated; however, residual effector activity has been observed for some immune complexes formed with non-glycosylated IgG-Fc due to multiple IgG-Fc/ligand interactions and resultant increased binding due to avidity (*79–84*). Whereas a therapeutic modality requires antibody effector function activation, 100 % oligosaccharide occupancy is a CQA; by the same token, if an aglycosylated IgG is to be employed, 0 % occupancy is a CQA. Initially, CHO, NS0, and Sp2/0 cells were used for the production of mAbs (*84–87*). These cell lines produce predominantly G0F IgG-Fc glycoforms with relatively low levels of galactosylated and non-fucosylated IgG-Fc relative to normal polyclonal IgG-Fc; CHO cells may add N-acetylneuraminic acid residues but in α(2–3) linkage rather than the α(2–6) linkage present in humans. In addition, these cell lines may add sugars that are not present in normal serum-derived IgG and can be immunogenic in humans. A particular concern is the addition, by NS0 and Sp2/0 cells, of galactose in α(1–3) linkage to galactose linked β(1–4) to the N-acetylglucosamine residues (*87–90*). Humans and higher primates do not have a functional gene encoding the transferase that adds galactose in α(1–3) linkage, however, due to continual environmental exposure to the gal α(1–3) gal epitope. For example, in red meats, humans develop IgG antibody that is specific for this antigen (*86–90*). The (gal α(1–3) gal) epitope is widely expressed on hamster cells, and it has recently been reported that some derived CHO cell lines are capable of (gal α(1–3) gal) addition. Similarly, CHO, NS0, and Sp2/0 cells may add an N-glycolylneuraminic acid, in α(2–3) linkage, that is not present in humans and may also be immunogenic (*84–90*). A significant population of normal human IgG-Fc bears a bisecting N-acetylglucosamine residue that is absent from IgG-Fc produced in CHO, NS0, or Sp2/0 cells. The generation of homogeneous IgG-Fc glycoforms *in vitro* has shown that the effector functions activated qualitatively and quantitatively differ between IgG subclasses and antibody glycoforms. It has not proved possible to manipulate culture medium conditions to generate predetermined mAb glycoform profiles; however, significant "tweaking" of the profile can be achieved during a production run (*91, 92*), and cellular engineering has been employed to enhance production of particular human IgG-Fc glycoforms, see below.

IgG-Fc Ligand Binding, Activation, and Modulation

Cellular IgG-Fc Receptors

Two distinct functions for IgG-Fc receptors may be distinguished: (1) to bind antibody–antigen complexes and initiate effector functions leading to their removal and destruction and (2) to mediate transport across epithelial membranes (i.e., transcytosis).

IgG-Fc Receptors (FcγR) Mediating Antigen Clearance

Three types/classes of membrane-bound human FcγR (FcγRI [CD64], FcγRII [CD32], FcγRIII [CD16]) and six subtypes (FcγRI, FcγRIIA, FcγRIIB1, FcγRIIB2, FcγRIIC, FcγRIIIA, and FcγRIIIB) have been defined by immunochemical, biochemical, and gene sequencing studies (*93–97*). These receptors are variously and constitutively expressed on a wide range of leucocytes, and expression may be up regulated and/or induced by cytokines generated and released within an inflammatory response. The effector mechanisms activated are diverse and include "killing" and removal (e.g., phagocytosis, respiratory burst, and cytolysis), accessory functions such as the enhancement of antigen presentation by dendritic cells, and the down-regulation of growth and differentiation of lymphocytes. It is evident, therefore, that FcγRs play an essential role in the induction, establishment, and resolution of protective immune responses.

Multiple parameters will determine the structure and biological activities of immune complexes formed within a polyclonal antibody response: (1) valency, (2) average affinity/avidity of the antibody population, (3) isotype profile, (4) IgG-Fc glycoforms, (5) valency or epitope density of the antigen, (6) density of cell surface effector ligands (e.g., FcγR), (7) cumulative valency when multiple ligands are engaged (e.g., FcγR and complement receptors), and (8) proportions of each antibody isotype within a polyclonal response (*12, 17, 71–73, 79–84, 93–97*).

The FcγRI receptor is constitutively expressed on mononuclear phagocytes and dendritic cells; however, expression can be up-regulated and/or induced by the action of cytokines. FcγRIIa is the most widely expressed FcγR and is found on most hemopoietic cells. Polymorphic variants of FcγIIa are identified by the presence of histidine (FcγRIIa-131H) or arginine (FcγRIIa-131R) at amino acid residue 131. The higher affinity of the FcγRIIa-131H form for IgG2 results in differing cellular responses to engagement by IgG2 immune complexes (*12, 93–97*); higher phagocytic capacity for *Streptococcus pneumoniae* opsonised with IgG2 antibody was observed for neutrophils of donors homozygous for FcγRIIa-131H than for FcγIIa-131R. The FcγRIIb receptor is expressed on B-lymphocytes and monocytes, and ligation of this receptor results in growth and differentiation inhibition (*12, 93–96*).

Initially the FcγRIIIa receptor was reported to bind and be activated by IgG1 and IgG3 only; however, recognition of a polymorphism in the receptor and the differential influence of IgG glycoforms has radically changed our understanding, with important clinical consequences. The avidity of binding of IgG differs between the FcγRIIIa-158V and FcγRIIIa-158F polymorphic variants (*97*). It was demonstrated *in vitro* that IgG1 antibody is more efficient at mediating ADCC through homozygous FcγRIIIa-158V bearing cells than homozygous FcγRIIIa-158F or heterozygous FcγRIIIa-158V/FcγRIIIa-158F cells (*97–99*). Similar differences in ADCC efficacy might pertain *in vivo* because when exposed to Rituxan, more favourable responses were reported for patients diagnosed with systemic lupus erythematosus or leukaemia and who were homozygous for FcγRIIIa-158V than for homozygous FcγRIIIa-158F patients (*97, 98*). Similarly, FcγRIIIa polymorphisms were shown to influence the response of Crohn's disease patients to infliximab (*99*) and red blood cell clearance by anti-D antibody (*100*).

All FcγR are transmembrane molecules except FcγRIIIb, which is glycosylphosphatidylinositol (GPI)-anchored within the membrane of neutrophils. FcγRI and FcγRIIIa are members of the multi-chain immune recognition receptor (MIRR) family and are present in the membrane as hetero-oligomeric complexes comprised of an α and a γ chain. An IgG–antigen complex binds the α chain to initiate signalling through the γ chain; the FcγRIIIa α chain of natural killer (NK) cells is also associated with a signalling ζ chain. FcγRIIa and FcγRIIb molecules are comprised of an α chain only (*93–96*). The FcγR α chains show a high degree of sequence homology in their extracellular domains (70-98%) but differ significantly in their cytoplasmic domains. The cytoplasmic domains of γ chains and the FcγRIIa α chain express the immunoreceptor tyrosine-based activation motif (ITAM) that is involved in the early stages of intracellular signal generation. By contrast, the FcγRIIb receptor α chain expresses an immunoreceptor tyrosine-based inhibition motif (ITIM) (*12–14, 93–96*). Cellular activation may be dependent on the balance between the relative levels of expression of these two isoforms and, hence, the balance of signals generated through the ITAM and ITIM (*93–96*).

FcγR Binding Sites on IgG

The crystal structure of IgG-Fc in complex with soluble recombinant FcγRIIIb, FcγRIIIa and FcγRIIa reveals direct involvement of the lower hinge and hinge-proximal C_H2 domain residues (*12, 101–106*). One primary publication suggested a possible contribution of the IgG-Fc N-acetylglucosamine residue to binding (*101*), whilst another primary publications held that there is no direct contact (*102, 104*); in a subsequent review, the authors of the latter publication stated that the oligosaccharides contributes ~100 $Å^2$ to the contact interface (*105*); this conclusion resulted from refinement of the previously obtained crystal data (P. Sun, personal communication). These investigators also demonstrated that whilst the binding of deglycosylated IgG-Fc to the *Escherichia coli* (*E. coli*)-derived aglycosylated FcγRIIIb was undetectable, the binding of deglycosylated whole IgG was only decreased 10–15-fold (*105*). This serves to remind us to exercise caution when tempted to extrapolate from *in vitro* experimental data to *in vivo* biological outcomes. Analysis of the complex formed between IgG-Fc and recombinant glycosylated FcγRIIIa has confirmed the general features of the interaction but has also revealed a critical role for glycosylation of FcγRIIIa (*103, 106*), see below. Thus, the IgG-Fc-FcγRIIIa interaction is significantly influenced by the glycoform of each component. The involvement of both heavy chains in the formation of an asymmetric binding site provides a structural explanation for an essential requirement: that the IgG should be univalent for the FcγR; if monomeric IgG were divalent, it could cross-link cellular receptors and hence constantly activate inflammatory reactions.

FcRn: Catabolism and Transcytosis

Transcytosis

The FcRn receptor was first identified from studies of the transport of IgG across the gut of newborn rats and designated the neonatal Fc receptor, hence FcRn; subsequently, the human homologue was identified in human placenta and shown to mediate transport of IgG from mother to foetus (*107–110*). The interaction site on IgG-Fc is at the C_H2/C_H3 interface, and the C_H3 sequence -H-N-H-Y-H- (Eu: 433-436) is of functional significance because titration of these histidine residues accounts for the observed binding of IgG to FcγRn at pH 6.0-6.5 and its release at pH 7.0-7.5 (*107–110*). The interaction of IgG-Fc with FcRn appears not to be influenced by the natural IgG-Fc glycoform profile, or indeed the presence or absence of oligosaccharides. On the contrary, the *N*-glycans in FcRn contribute to the steady-state membrane distribution and direction of IgG transport (*109–111*)

Each of the four human IgG subclasses are transported across the placenta, however, with differing facility. Cord-blood levels of IgG1 may be higher than in matched maternal blood, whilst IgG3 and IgG4 levels are equivalent. The level of IgG2 is ~80 % of the concentration in maternal blood (*112*). It is of interest to note that IgG3 is transferred with equal efficacy to IgG1, although it has a shorter half-life, suggesting that its interaction with FcRn in the environment of the placenta may be different from that in endosomes in the catabolic pathway. During pregnancy, the level of galactosylation of maternal IgG increases, and there is preferential transport of galactosylated IgG across the placenta (*63, 64, 68, 112*). This provides circumstantial evidence to suggest that the affinity of IgG for FcγRn may differ between glycoforms under conditions operative at the interface between the mother and the placenta.

The potential protein therapeutics having short half-lives (e.g., cytokines) are being generated as fusion proteins with IgG-Fc to extend the half-lives. A further development opens a new route for administration because it has been shown that FcRn is expressed in the central and upper airways of the lung and that drug-IgG-Fc fusion proteins delivered to these sites can be transported by transcytosis to the systemic circulation. This is an exciting development with considerable promise and significance (*113–115*).

Catabolism

The catabolic pathway of human IgG antibodies is also mediated by FcRn that is expressed by many tissues (*12–14, 116–119*). FcRn-expressing cells take up IgG within pinocytotic vesicles, resulting in the formation of vacuoles. Subsequent lowering of pH results in saturation binding to FcRn and protection of bound IgG from cleavage by enzymes present in the vacuole; unbound IgG is degraded. When the membrane of the vacuole is re-cycled to the cellular membrane, the IgG-FcRn complex is exposed to pH 7.2 and released into the extra-vascular fluid. The catabolic half-life of human IgG1, 2, and 4 is ~21 days, whilst for IgG3, it is ~7 days; the IgG3 data was generated for IgG3 molecules of G3m(5*) and G3m(21) allotype in which arginine is present at residue 435, in contrast to the histidine residue that is present in IgG1, IgG2, and IgG4 (*12, 16, 116–119*). It has been reported that replacement of the IgG3 arginine 435 residue by histidine increases the affinity for FcRn, and consequently, the half-life (*119*). The IgG allotype G3m(15,16), present within Mongoloid populations, has a histidine residue at position 435 and higher binding affinity for FcRn (*110, 119*). The catabolic half-life of IgG mediated through FcRn does not appear to be influenced by IgG-Fc glycoform (*12–14, 111–114, 116*); however, it should be emphasised that only glycoforms of IgG bearing neutral oligosaccharides have been evaluated. It may be anticipated that sialylation could influence IgG-FcRn interactions because it would introduce a negative charge in the vicinity of the histidine residues involved in FcRn binding. Protein engineering has been successfully applied to increase the affinity of IgG-Fc binding to FcRn and, hence, increase the catabolic half-life (*117–119*). Prolongation of the half-life of an IgG therapeutic could translate into reduced frequency of dosing and attendance at clinic, thus reducing the cost of treatment. There is evidence that glycoproteins expressing terminal N-acetylglucosamine residues may be cleared through the mannose receptor and accounts for the enhanced clearance of the IgG-Fc-TNF (tumour necrosis factor) receptor fusion protein therapeutic (Lenercept) having exposed terminal N-acetylglucosamine residues (*120, 121*).

Complement Activation

Classical Pathway: C1q/C1 Binding and Activation

Activation of complement through the classical pathway results in a cascade of enzymatic reactions with subsequent cleavage of downstream complement components with amplification at each step and the generation of fragments that (1) bind to an immune complex; (2) recruit leucocytes to augment an inflammatory response; (3) form a multimeric "membrane attack complex" that inserts into cellular and bacteria membranes to generate a "pore" that allows the ingress of water and consequent lysis (*12–14, 122–124*). Leucocytes expressing FcγR may also express receptors for complement fragments bound to an immune complex, thus enhancing opsonisation (*12–14, 122–124*).

Glycosylation of IgG-Fc is essential for C1 binding and activation, and immune complexes incorporating IgG1 and/or IgG3 antibody are highly active, whilst only IgG2 complexes formed in antigen excess may be active. There has been a consensus that IgG4 does not activate the classical pathway (*12–14, 125–130*); however, ordered IgG4 hexameric complexes have been reported to activate C1 (*131*). Activity can be modulated by protein- and glyco-engineering, and a hybrid IgG1/IgG3 molecule has been shown to exhibit enhanced activity relative to either IgG1 or IgG3 alone (*125*); this antibody format is available commercially as Complegent. Protein engineering studies suggest that the interaction site of human IgG1 for C1 is localised to the hinge-proximal region of the C_H2 domain (*12–14, 126, 130*). This proposal is further supported by the demonstration that replacement of the Pro 331 residue of IgG4 by serine converts it to a molecule that can activate C1; proline 331 is localised at the junction of the b6 bend and the fy3 β-strand that is topographically proximal to hinge region (*12–14, 129*). Extensive studies of mutant chimeric human IgG3 proteins have established that the efficiency of C1 activation is not directly determined by the length of the hinge region but that at least one inter-heavy chain disulphide bridge is required (*12–14, 45*).

Role of IgG Glycoforms in Recognition by Cellular FγRs

Because glycosylation of IgG-Fc is essential to recognition and activation of effector ligands, quantitative and/or qualitative functional differences between glycoforms might be anticipated. This was confirmed in studies employing recombinant mAbs glyco-engineered *in vitro* and led to the engineering of cell lines to produce antibodies having a predetermined glycoform profile (*12, 71–73, 132–136*). These research finding lead one to speculate whether the immune system responds to pathogens by production of both an optimal antibody isotype and glycoform profile. It should be noted that the impact of antibody glycoform on ligand binding/activation may differ depending on the assay format adopted. Thus, measurements of monomeric IgG antibody-binding affinity for a ligand immobilised on a solid phase may deliver a very different assessment to its ability to sensitise a cell line for killing by peripheral blood mononuclear cells (PBMCs) or enhancement of clinical efficacy *in vivo*. The high sensitivity of current analytical protocols, particularly mass spectrometry, is now providing the tools to determine the glycoform profile of specific antibody populations, and individual glycoform profiles have been reported for antiplatelet auto-antibodies and anti-citrullinated peptide antibodies (*52, 67*); it remains to be determined whether these differences relate to disease activity.

The Influence of Fucose and Bisecting N-Acetylglucosamine on IgG-Fc Activities

Increased ADCC was reported for antibodies lacking the presence of fucose, produced in Lec13 CHO cells (*132*), and for antibodies expressing bisecting N-acetylglucosamine, produced in CHO cells transfected with the human β 1,4-N-acetylglucosaminyltransferase III (GnTIII) gene (*133–135*). It was further demonstrated that the presence of bisecting N-acetylglucosamine inhibits the endogenous α(1,6)-fucosyltransferase and the addition of fucose (*136*). This glycoform, bearing bisecting N-acetylglucosamine with the absence of fucose, is a minor component (<3 %) of oligosaccharides released from normal polyclonal human IgG-Fc; however, it may be a predominant glycoform produced by an individual plasma cell clone (*55, 62*).

The Influence of Galactosylation on IgG-Fc Activities

The extent of IgG-Fc galactosylation is a major source of glycoform heterogeneity in health and disease. Compared to levels of galactosylation observed for young adults, there is a decline with continuing ageing; there is also a small but significant gender difference (*137, 138*). An increase in IgG-Fc galactosylation occurs over the course of normal pregnancy, with levels returning to the adult norm following parturition (*63, 64*). Hypogalactosylation of IgG-Fc is reported for a number of inflammatory states associated with autoimmune diseases (*65–68, 139–141*). The extent of IgG-Fc galactosylation observed for monoclonal human myeloma IgG proteins is highly variable, indicating that the level of IgG-Fc galactosylation is a clonal property (*61, 62, 142*). The antibody products of CHO, Sp2/0 and NS0 cell lines used in commercial production of recombinant antibody are generally highly fucosylated but hypogalactosylated, relative to normal polyclonal human IgG (*69, 74, 143–145*); it is necessary therefore, to consider the possible impact of differential IgG-Fc galactosylation on functional activity.

Numerous studies have probed the influence of the presence or absence of galactose residues on IgG-Fc structure and function. A NMR study of galactosylated [G2F]2 and agalactosylated [G0F]2 glycoforms of IgG-Fc reported the mobility of the glycan to be comparable to that of the backbone polypeptide chain, with the exception of the galactose residue on the α(1−3) arm, which was highly mobile. It was concluded that agalactosylation does not induce any significant change in glycan mobility or protein conformation (*146–148*). This report is consistent with FcγR binding and stability studies showing minimal differences between [G0F]2 and [G2F]2 glycoforms (*71–73*) and crystal structures of a series of truncated glycoforms of IgG-Fc (*73*). The NMR study also probed changes in local environments on the binding of soluble recombinant FcγRIII to [G2F]2 and [G0F]2 glycoforms of IgG1-Fc and reported chemical shift differences > 0.2 ppm for K248 and V308 residues (*146*); this is a very localised change distant from the interaction site for the FcγRIIIa moiety. The finding of a changed environment for these residues is interesting because from the crystal structure, they were not predicted to make contacts with the α(1-6) arm galactose residue; small perturbations for the oligosaccharide contact residues K246, D249, T256 were also observed. A comparison of hydrogen−deuterium exchange for [G2F]2 and [G0F]2 IgG-Fc glycoforms revealed altered conformation within the peptide sequence 242 – 254 (*148*).

The possible consequences for hypogalactosylated recombinant antibodies on *in vivo* activity have been extrapolated from *in vitro* assays and animal experiments. Removal of terminal galactose residues from Campath-1H was shown to reduce classical complement activation but to be without effect on FcγR-mediated functions (*150*). Similarly, the ability of rituximab to kill tumour cells by the classical complement route has been shown to be maximal for the [G2F]2 glycoform in comparison to the [G0F]2 glycoform (*145*). The product that gained licensing approval was comprised of ~25 % of the G1F oligosaccharide; therefore, regulatory authorities required that galactosylation of the manufactured product be controlled to within a few percent of this value. In the absence of galactose, the terminal sugar residue is N-acetylglucosamine, which may be accessible to both the mannose receptor and/or mannan-binding lectin.

Sialylation of IgG-Fc Oligosaccharides

A minority of oligosaccharides released from polyclonal IgG-Fc are sialylated, whilst ~70 % bear one or two galactose residues (*51–54, 79–84, 149*). The paucity of sialylation is presumed to reflect restricted access of terminal galactose residues for the α(2−6) sialyltransferase enzyme due to the generation of a "closed" IgG-Fc protein conformation rather than due to any deficit in the sialylation machinery. This conclusion is supported by the finding that when both IgG-Fc and IgG-Fab are glycosylated, the latter bears highly galactosylated and sialylated structures, demonstrating that the glycosylation machinery is fully functional (*12, 55*). The presence or absence of terminal galactose and/or sialic acid residues does not influence IgG catabolism because it is not catabolised in the liver via the asialoglycoprotein receptor (ASGPR) but in multiple cell types that express the FcRn receptor. The balance between structure and accessibility is well illustrated for a panel of IgG-Fcs in which individual amino acid residues making contacts with the oligosaccharide were replaced by alanine. In each case, hypergalactosylated and highly sialylated glycoforms resulted, suggesting some relaxation of structure allowing access to glycosyl transferases (*71–73*, *150*)

Recent studies suggest that sialylated human IgG-Fc may inhibit activation of inflammatory cascades. Following binding to the lectin receptor SIGN-R1, in the mouse, or DC-SIGN, in humans, expression of the inhibitory receptor FγRIIb on inflammatory cells is up-regulated, attenuating auto-antibody-initiated inflammation (*151*, *152*). However, it has been asserted that caution should be exercised when extrapolating from mouse models to humans because the tissue distributions of SIGN-1 and DC-SIGN differ and because anti-inflammatory activity could be demonstrated for intact IgG and F(ab')$_2$ fragments (*153–155*).

Quaternary Structure of Fab

Structures for numerous antigen-specific Fab fragments have been determined by x-ray crystallography, both alone and in complex with antigen (*15*, *156*). Early studies of Fab fragments binding small molecules (haptens) led to the lock-and-key model in which the antigen bound within a pocket; however, studies of macromolecular antigens have shown that both paratopes and epitopes may be comprised of relatively flat surfaces (*15*, *156*). Whilst light and heavy chain CDRs are seen to contribute to epitope binding for all specificities, it is not the case that all CDRs contribute equally to a given paratope. Comparison of the V_H and V_L sequences of a given Fab with that encoded within the precursor germline gene informs of the contribution of somatic hypermutation to specificity and/or affinity both within and outside of the paratope. These studies have defined the contribution of CDRs and "framework" sequences to antigen binding and led to the development of informed protocols for humanisation of mouse antibody V regions for the generation of therapeutic antibodies (*157–159*). Mapping of the surface topology of Fab regions identifies the degree of exposure of amino acid side chains and can contribute to further informed protein engineering to confer advantageous properties (e.g., solubility) (*157–159*).

Comparison of crystal structures of Fab fragments in the free and antigen-bound form show that conformational change may occur within V_H and V_L domains on antigen binding (*156*, *160–162*). This provides further understanding of epitope recognition and binding and may reveal residues underlying the paratope, the vernier zone, that are essential to its architecture, mobility and, hence, specificity and/or affinity (*163*, *164*). The junction of the V_H/C_H1 and V_L/C_L, referred to as the "switch" residues, is characterised by a change in direction of the polypeptide chains, referred to as the "elbow angle". The elbow angle is characteristic for a given Fab but can vary widely among Fabs; lambda light chains appear to be compatible with larger elbow angles than kappa light chains (*165*). There is evidence that conformational changes resulting from antigen binding can be transmitted to the C_H1/C_L domain (*160*); also, the reciprocal finding is that differences in C_H1 structure can influence antigen-binding affinity (*166*).

An understanding of the structure and function of the Fab region of antibodies facilitated genetic engineering to progressively "humanise" antibodies raised in mice (*157–159*). The engineered V_H and V_L gene sequences are ligated to the constant region gene sequences of heavy and light chains, respectively, to generate humanised antibodies. In practice, such "humanized" V_H and V_L domains usually resulted in reduced affinity and/or specificity such that some mouse residues had to be re-introduced, with consequent potential immunogenicity. Libraries of human V_H and V_L gene sequences have been generated from human peripheral blood lymphocytes using polymerase chain reaction (PCR) protocols, and their protein products expressed in phage display libraries, allowing for selection of combinations of human V_H and V_L sequences with specificity for a selected target; the V_H and V_L sequences can subsequently be expressed with selected human constant regions (*157–159, 163–165*). Fully human antibodies specific for selected human targets can be realised by immunisation of mice that are transgenic for the expression of human variable and constant region genes, the endogenous Ig genes having been inactivated (*166*). It may not be possible to dictate the human constant region expressed but subsequent engineering can rectify this.

Clinical experience demonstrated that "fully" human antibodies can be immunogenic, at least in a proportion of patients (*167, 168*). Initially, one may wonder why this should be so; on reflection, however, it may seem inevitable. The hallmark of an antibody is its specificity, not just for a particular target but also for a unique structural feature on that target—the epitope. This is achieved within a secondary immune response that is characterised by somatic hypermutation and selection. Thus, each antibody is a structurally unique molecule with a unique epitope-binding site (the paratope). Antibodies generated from phage display libraries or transgenic mice are unique to an individual—human or mouse—and may be perceived as "foreign" within an outbred population of unique individual recipients (i.e., patients). Antibody therapeutics are manufactured in xenogeneic tissue (e.g., Chinese hamster, mouse) that may yield product not having the required human-type co- and post-translational modifications and/or add having added nonhuman co- and post-translational modifications (*12–16*).

Human Antibody Isotypes Other Than IgG

This review has focused on antibody therapeutics; consequently, discussion of structure–function relationships has been restricted to the IgG isotypes; however, our understanding of the structure and function of the IgA (*5, 169–171*), IgM (*5, 172, 173*) and IgE (*5, 174*) isotypes is progressing and therapeutic applications are on the horizon or in the pipeline, if not licensed (*175–178*).

Concluding Remarks

It is salutary to contemplate the finesse of structural changes induced by conservative amino acid replacements (G/A235; F/A243) (*12, 129, 135, 150, 179*) and/or the effects that the presence or absence of a fucose and/or bisecting N-acetylglucosamine sugar residue can have on the functional activity of the human IgG-Fc (*132–136*). It is evident, therefore, that the conformation of the IgG molecule is a CQA that is relatively robust while being amenable to protein engineering. It is essential, therefore, that multiple orthogonal techniques should be applied to determine structural parameters; both industry and academia will be best served by having access to a reference standard that has been characterised employing relevant "state of the art" techniques. In this review, I have attempted to present the current consensus of understanding of IgG structure and function; however, it is far from complete. It is interesting to note that disparate ligands may bind to the Fc through common amino acid residue contacts within the hinge-proximal region (for FcγR and C1q) and the C_H2/C_H3 interface (for FcRn, SpA, SpG, RFs, and IgG-Fc-like receptors encoded within the genomes of some viruses). The presence of sialic acid might further influence Fc–ligand interactions at this interface. A rationalisation for the topography of ligand binding sites may be the functional necessity for circulating IgG to be monovalent for FcγR's and C1q to prevent continuous cellular activation while providing opportunity for divalency at the C_H2/C_H3 interface. The influence of the IgG-Fc glycoform on functional activity may be exploited to generate homogeneous glycoforms selected to express a predetermined functional profile considered optimal for a given disease indication. It is important to note that each glycoform is represented within normal polyclonal IgG-Fc; therefore, they do not have immunogenic potential. Many innovative studies have explored engineering of the protein moiety to selectively enhance biologic activities (*12*, *129*, *135*, *150*, *179*); however, these are mutant forms of IgG (i.e., non-self, which may enhance immunogenicity). This may not be an issue when treating patients for cancer because they may be immune-suppressed; however, it is a concern for long-term treatment of chronic diseases.

References

1. Kohler, G.; Milstein, C. *Nature* **1975**, *256*, 495–497.
2. May, W.; Parris, C.; Beck, J.; Fassett, R.; Greenberg, F.; Guenther, G.; Kramer, G.; Wise, S.; Gils, T.; Bolbert, R.; MacDonald, B. *Definitions of Terms and Modes Used at NIST for Value-Assignment of Reference Materials for Chemical Measurements*; NIST Special Publication 260-136; National Institute of Standards and Technology: Gaithersburg, MD, 2000; pp 1–18.
3. *Intoduction to Biological and Small Molecule Drug Research: Theory and Case Studies*; Ganellin, C., Jefferis, R., Roberts, R., Eds.; Academic Press: Oxford, U.K., 2013.
4. van Hartingsveldt, B. Small Molecules vs Biologics: Different Early Development. http://www.thepharmaceuticalconference.be/_docs/2013/presentations/BPC%202013%20-%20Bart%20van%20Hartingsveldt.pdf (accessed January 4, 2014).
5. Strohl, R.; Strohl, L. *Therapeutic Antibody Enginering: Current and Future Advances Driving the Strongest Growth Area in the Parmaceutical Industry*; Woodhead Publishing, Ltd: Cambridge, U.K., 2012.
6. FDA Guidance for Industry: Scientific Considerations in Demonstrating Biosimilarity to a Reference Product. http://www.fda.gov/downloads/Drugs/

GuidanceComplianceRegulatoryInformation/Guidances/UCM291128.pdf (accessed 1-4-2014).
7. EMA Guideline on Development, Production, Characterisation and Specifications for Monoclonal Antibodies and Related Products. http://www.ema.europa.eu/docs/en_GB/document_library/Scientific_guideline/2009/09/WC500003074.pdf (accessed January 4, 2014).
8. Hospira Press Release. http://phx.corporate-ir.net/phoenix.zhtml?c=175550&p=irol-newsArticle&ID=1853480&highlight (accessed January 4, 2014).
9. Beck, A.; Reichert, J. M. *mAbs* **2013**, *5*, 621–623.
10. Iwasaki, A.; Medzhitov, R. *Science* **2010**, *327*, 291–295.
11. Gardy, J. L.; Lynn, D. J.; Brinkman, F. S.; Hancock, R. E. *Trends Immunol.* **2009**, *30*, 249–262.
12. Jefferis, R. *Arch. Biochem. Biophys.* **2012**, *526*, 159–166.
13. Nezlin, R.; Ghetie, V. *Adv. Immunol.* **2004**, *82*, 155–215.
14. Schroeder, H. W., Jr.; Cavacini, L. *J. Allergy Clin. Immunol.* **2010**, *125*, S41–52.
15. Narciso, J. E.; Uy, I. D.; Cabang, A. B.; Chavez, J. F.; Pablo, J. L.; Padilla-Concepcion, G. P.; Padlan, E. A. *N. Biotechnol.* **2011**, *28*, 435–447.
16. Jefferis, R.; Lefranc, M.-P. *mAbs* **2009**, *1*, 332–338.
17. Jefferis, R. *mAbs* **2011**, *3*, 503–504.
18. Lux, A.; Yu, X.; Scanlan, C. N.; Nimmerjahn, F. *J. Immunol.* **2013**, *190*, 4315–4323.
19. Edelman, G. M.; Cunningham, B. A.; Gall, W. E.; Gottlieb, P. D.; Rutishauser, U.; Waxdal, M. J. *J. Immunol.* **2004**, *173*, 5335–5342.
20. Deisenhofer, J. *Biochemistry* **1981**, *20*, 2361–2370.
21. Saphire, E. O.; Stanfield, R. L.; Crispin, M. D. M.; Parren, P.; Rudd, P. M.; Dwek, R. A.; Burton, D. R.; Wilson, I. A. *J. Mol. Biol.* **2002**, *319*, 9–18.
22. Liu, Y. D.; Wang, T.; Chou, R.; Chen, L.; Kannan, G.; Stevenson, R.; Goetze, A. M.; Jiang, X. G.; Huang, G.; Dillon, T. M.; Flynn, G. C. *Mol. Immunol.* **2013**, *54*, 217–226.
23. Grubb, R. *Acta Pathol. Microbiol. Scand.* **1956**, *39*, 195–197.
24. Lefranc, M. P.; Lefranc, G. *Methods. Mol. Biol.* **2012**, *882*, 635–680.
25. Jefferis, R. *Immunol. Today* **1993**, *14*, 119–121.
26. International Union of Immunological Sciences. *Eur. J. Biochem.* **1974**, *45*, 5–6.
27. WHO *Eur. J. Immunol.* **1976**, *6*, 59.
28. Lefranc, M. Imgt Repertoire (Ig and Tr). Isotypes: Human (Homo Sapiens) Iglc. http://www.imgt.org/IMGTrepertoire/Proteins/isotypes/human/IGL/IGLC/Hu_IGLCisotypes.html (accessed January 4, 2014).
29. Dard, P.; Lefranc, M. P.; Osipova, L.; Sanchez-Mazas, A. *Eur. J. Human Gen.* **2001**, *9*, 765–772.
30. Dechavanne, C.; Guillonneau, F.; Chiappetta, G.; Sago, L.; Levy, P.; Salnot, V.; Guitard, E.; Ehrenmann, F.; Broussard, C.; Chafey, P.; Le Port, A.; Vinh, J.; Mayeux, P.; Dugoujon, J. M.; Lefranc, M. P.; Migot-Nabias, F. *PLoS One* **2012**, *7*, e46097.

31. Lefranc, G. Imgt Repertoire (Ig and Tr). Gm Allotypes and Gm Haplotypes http://www.imgt.org/IMGTrepertoire/Proteins/allotypes/human/IGH/IGHC/Hu_IGHCallotypes1.html (accessed Janaury 4, 2014).
32. Hougs, L.; Garred, P.; Kawasaki, T.; Kawasaki, N.; Svejgaard, A.; Barington, T. *Tissue Antigens* **2003**, *61*, 231–239.
33. Lefranc, G. Imgt Repertoire (Ig and Tr). Allotypes: Human (Homo Sapiens) Igkc. http://www.imgt.org/IMGTrepertoire/Proteins/allotypes/human/IGK/IGKC/Hu_IGKCallotypes.html 01.04.2014 (accessed January 4, 2014).
34. Dard, P.; Huck, S.; Frippiat, J. P.; Lefranc, G.; Langaney, A.; Lefranc, M. P.; SanchezMazas, A. *Hum. Genet.* **1997**, *99*, 138–141.
35. Labrijn, A. F.; Rispens, T.; Meesters, J.; Rose, R. J.; den Bleker, T. H.; Loverix, S.; van den Bremer, E. T.; Neijssen, J.; Vink, T.; Lasters, I.; Aalberse, R. C.; Heck, A. J.; van de Winkel, J. G.; Schuurman, J.; Parren, P. W. *J. Immunol.* **2011**, *187*, 3238–3246.
36. Carter, P.; Presta, L.; Gorman, C. M.; Ridgway, J. B. B.; Henner, D.; Wong, W. L. T.; Rowland, A. M.; Kotts, C.; Carver, M. E.; Shepard, H. M. *Proc. Nat. Acad. Sci. U.S.A.* **1992**, *89*, 4285–4289.
37. Gorman, S. D.; Clark, M. R. *Semin. Immunol.* **1990**, *2*, 457–466.
38. Harris, L. J.; Skaletsky, E.; McPherson, A. *J. Mol. Biol.* **1998**, *275*, 861–872.
39. Harris, L. J.; Larson, S. B.; Skaletsky, E.; McPherson, A. *Immunol. Rev.* **1998**, *163*, 35–43.
40. Gregory, L.; Davis, K. G.; Sheth, B.; Boyd, J.; Jefferis, R.; Nave, C.; Burton, D. R. *Mol. Immunol.* **1987**, *24*, 821–829.
41. Eryilmaz, E.; Janda, A.; Kim, J.; Cordero, R. J.; Cowburn, D.; Casadevall, A. *Mol. Immunol.* **2013**, *56*, 588–598.
42. Borrok, M. J.; Jung, S. T.; Kang, T. H.; Monzingo, A. F.; Georgiou, G. *ACS Chem. Biol.* **2012**, *7*, 1596–1602.
43. Yamaguchi, Y.; Nishimura, M.; Nagano, M.; Yagi, H.; Sasakawa, H.; Uchida, K.; Shitara, K.; Kato, K. *Biochim. Biophys. Acta, Gen. Subj.* **2006**, *1760*, 693–700.
44. Barb, A. W.; Prestegard, J. H. *Nat. Chem. Biol.* **2011**, *7*, 147–153.
45. Brekke, O. H.; Michaelsen, T. E.; Sandlie, I. *Immunol. Today* **1995**, *16*, 85–90.
46. Padlan, E., X-ray Diffraction Studies of Antibody Constant Regions. In *Fc Receptors and the Action of Antibodies*; Metzger, H., Ed.; American Society for Microbiology: Washington, DC, 1990; pp 12−30.
47. Jefferis, R.; Lund, J.; Pound, J. D. *Immunol. Rev.* **1998**, *163*, 59–76.
48. Ozbabacan, S. E. A.; Engin, H. B.; Gursoy, A.; Keskin, O. *Protein Eng., Des. Sel.* **2011**, *24*, 635–648.
49. Sondermann, P.; Huber, R.; Oosthuizen, V.; Jacob, U. *Nature* **2000**, *406*, 267–273.
50. Radaev, S.; Motyka, S.; Fridman, W. H.; Sautes-Fridman, C.; Sun, P. D. *J. Biol. Chem.* **2001**, *276*, 16469–16477.
51. Reusch, D.; Haberger, M.; Selman, M. H.; Bulau, P.; Deelder, A. M.; Wuhrer, M.; Engler, N. *Anal. Biochem.* **2013**, *432*, 82–89.
52. Huhn, C.; Selman, M. H. J.; Ruhaak, L. R.; Deelder, A. M.; Wuhrer, M. *Proteomics* **2009**, *9*, 882–913.

53. Flynn, G. C.; Chen, X.; Liu, Y. D.; Shah, B.; Zhang, Z. *Mol. Immunol.* **2010**, *47*, 2074–2082.
54. Anumula, K. R. *J. Immunol. Methods* **2012**, *382*, 167–176.
55. Mimura, Y.; Ashton, P. R.; Takahashi, N.; Harvey, D. J.; Jefferis, R. *J. Immunol. Meth.* **2007**, *326*, 116–126.
56. Symbol and Text Nomenclature for Representation of Glycan Stucture. Nomenclature Committee Consortium for Functional Glycomics. http://glycomics.scripps.edu/CFGnomenclature.pdf (accessed February 4, 2014).
57. GlycoBase 3.2. National Institute for Bioprocessing Research and Training (NIBRT). http://glycobase.nibrt.ie/glycobase/about.action (accessed February 4, 2014).
58. Masuda, K.; Kubota, T.; Kaneko, E.; Iida, S.; Wakitani, M.; Kobayashi-Natsume, Y.; Kubota, A.; Shitara, K.; Nakamura, K. *Mol. Immunol.* **2007**, *44*, 3122–3131.
59. Ferrara, C.; Brunker, P.; Suter, T.; Moser, S.; Puntener, U.; Umana, P. *Biotechnol. Bioeng.* **2006**, *93*, 851–861.
60. Kubota, T.; Niwa, R.; Satoh, M.; Akinaga, S.; Shitara, K.; Hanai, N. *Cancer Sci.* **2009**, *100*, 1566–1572.
61. Jefferis, R.; Lund, J.; Mizutani, H.; Nakagawa, H.; Kawazoe, Y.; Arata, Y.; Takahashi, N. *Biochem. J.* **1990**, *268*, 529–537.
62. Farooq, M.; Takahashi, N.; Arrol, H.; Drayson, M.; Jefferis, R. *Glycoconjugate J.* **1997**, *14*, 489–492.
63. Kibe, T.; Fujimoto, S.; Ishida, C.; Togari, H.; Wada, Y.; Okada, S.; Nakagawa, H.; Tsukamoto, Y.; Takahashi, N. *J. Clin. Biochem. Nutr.* **1996**, *21*, 57–63.
64. Alavi, A.; Arden, N.; Spector, T. D.; Axford, J. S. *J. Rheumatol.* **2000**, *27*, 1379–1385.
65. Ercan, A.; Cui, J.; Chatterton, D. E. W.; Deane, K. D.; Hazen, M. M.; Brintnell, W.; O'Donnell, C. I.; Derber, L. A.; Weinblatt, M. E.; Shadick, N. A.; Bell, D. A.; Cairns, E.; Solomon, D. H.; Holers, V. M.; Rudd, P. M.; Lee, D. M. *Arthritis Rheum.* **2010**, *62*, 2239–2248.
66. Holland, M.; Yagi, H.; Takahashi, N.; Kato, K.; Savage, C. O. S.; Goodall, D. M.; Jefferis, R. *Biochim. Biophys. Acta, Gen. Subj.* **2006**, *1760*, 669–677.
67. Scherer, H. U.; van der Woude, D.; Ioan-Facsinay, A.; el Bannoudi, H.; Trouw, L. A.; Wang, J.; Haeupl, T.; Burmester, G.-R.; Deelder, A. M.; Huizinga, T. W. J.; Wuhrer, M.; Toes, R. E. M. *Arthritis Rheum.* **2010**, *62*, 1620–1629.
68. Bondt, A.; Selman, M. H.; Deelder, A. M.; Hazes, J. M.; Willemsen, S. P.; Wuhrer, M.; Dolhain, R. J. *J. Proteome. Res.* **2013**, *12*, 4522–4531.
69. Pascoe, D. E.; Arnott, D.; Papoutsakis, E. T.; Miller, W. M.; Anderseni, D. C. *Biotechnol. Bioeng.* **2007**, *98*, 391–410.
70. Girardi, E.; Holdom, M. D.; Davies, A. M.; Sutton, B. J.; Beavil, A. J. *Biochem. J.* **2009**, *417*, 77–83.
71. Mimura, Y.; Church, S.; Ghirlando, R.; Ashton, P. R.; Dong, S.; Goodall, M.; Lund, J.; Jefferis, R. *Mol. Immunol.* **2000**, *37*, 697–706.

72. Mimura, Y.; Sondermann, P.; Ghirlando, R.; Lund, J.; Young, S. P.; Goodall, M.; Jefferis, R. *J. Biol. Chem.* **2001**, *276*, 45539–45547.
73. Krapp, S.; Mimura, Y.; Jefferis, R.; Huber, R.; Sondermann, P. *J. Mol. Biol.* **2003**, *325*, 979–989.
74. Schaefer, J. V.; Pluckthun, A. *J. Mol. Biol.* **2012**, *417*, 309–335.
75. Kameoka, D.; Ueda, T.; Imoto, T. *Appl. Biochem. Biotechnol.* **2011**, *164*, 642–654.
76. Zheng, K.; Bantog, C.; Bayer, R. *mAbs* **2011**, *3*, 568–576.
77. Allhorn, M.; Briceno, J. G.; Baudino, L.; Lood, C.; Olsson, M. L.; Izui, S.; Collin, M. *Blood* **2010**, *115*, 5080–5088.
78. Sjogren, J.; Struwe, W. B.; Cosgrave, E. F.; Rudd, P. M.; Stervander, M.; Allhorn, M.; Hollands, A.; Nizet, V.; Collin, M. *Biochem. J.* **2013**, *455*, 107–118.
79. Pound, J. D.; Lund, J.; Jefferis, R. *Mol. Immunol.* **1993**, *30*, 469–478.
80. Nesspor, T. C.; Raju, T. S.; Chin, C. N.; Vafa, O.; Brezski, R. J. *J. Mol. Recognit.* **2012**, *25*, 147–154.
81. Crispin, M. *Proc. Natl. Acad. Sci. U.S.A.* **2013**, *110*, 10059–10060.
82. Hristodorov, D.; Fischer, R.; Linden, L. *Mol. Biotechnol.* **2013**, *54*, 1056–1068.
83. Borrok, M. J.; Jung, S. T.; Kang, T. H.; Monzingo, A. F.; Georgiou, G. *ACS Chem. Biol.* **2012**, *7*, 1596–1602.
84. Jefferis, R. *Nat. Rev. Drug Disc.* **2009**, *8*, 226–234.
85. Li, F.; Vijayasankaran, N.; Shen, A. Y.; Kiss, R.; Amanullah, A. *mAbs* **2010**, *2*, 466–479.
86. Daguet, A.; Watier, H. *mAbs* **2011**, *3*, 417–421.
87. Ghaderi, D.; Zhang, M.; Hurtado-Ziola, N.; Varki, A. *Biotechnol. Genet. Eng. Rev.* **2012**, *28*, 147–175.
88. Galili, U. *Xenotransplantation* **2013**, *20*, 138–147.
89. Bosques, C. J.; Collins, B. E.; Meador, J. W., 3rd; Sarvaiya, H.; Murphy, J. L.; Dellorusso, G.; Bulik, D. A.; Hsu, I. H.; Washburn, N.; Sipsey, S. F.; Myette, J. R.; Raman, R.; Shriver, Z.; Sasisekharan, R.; Venkataraman, G. *Nat. Biotechnol.* **2010**, *28*, 1153–1156.
90. Maeda, E.; Kita, S.; Kinoshita, M.; Urakami, K.; Hayakawa, T.; Kakehi, K. *Anal. Chem.* **2012**, *84*, 2373–2379.
91. Yu, M.; Hu, Z.; Pacis, E.; Vijayasankaran, N.; Shen, A.; Li, F. *Biotechnol. Bioeng.* **2011**, *108*, 1078–1088.
92. Pacis, E.; Yu, M.; Autsen, J.; Bayer, R.; Li, F. *Biotechnol. Bioeng.* **2011**.
93. Nimmerjahn, F.; Ravetch, J. V., Fc Gamma Rs in Health and Disease. In *Negative Co-Receptors and Ligands*; Ahmed, R.; Honjo, T., Eds.; Springer: Berlin, 2011; Vol. 350, pp 105−125.
94. Lux, A.; Yu, X.; Scanlan, C. N.; Nimmerjahn, F. *J. Immunol.* **2013**, *190*, 4315–4323.
95. van de Winkel, J. G. J. *Immunol. Lett.* **2010**, *128*, 4–5.
96. Anthony, R. M.; Wermeling, F.; Ravetch, J. V. *Ann. N. Y. Acad. Sci.* **2012**, *1253*, 170–180.
97. Cartron, G.; Dacheux, L.; Salles, G.; Solal-Celigny, P.; Bardos, P.; Colombat, P.; Watier, H. *Blood* **2002**, *99*, 754–758.

98. Anolik, J. H.; Campbell, D.; Felgar, R. E.; Young, F.; Sanz, I.; Rosenblatt, J.; Looney, R. J. *Arthritis Rheum.* **2003**, *48*, 455–459.
99. Louis, E.; El Ghoul, Z.; Vermeire, S.; Dall'Ozzo, S.; Rutgeerts, P.; Paintaud, G.; Belaiche, J.; De Vos, M.; Van Gossum, A.; Colombel, J. F.; Watier, H. *Aliment. Pharmacol. Ther.* **2004**, *19*, 511–519.
100. Miescher, S.; Spycher, M. O.; Amstutz, H.; de Haas, M.; Kleijer, M.; Kalus, U. J.; Radtke, H.; Hubsch, A.; Andresen, I.; Martin, R. M.; Bichler, J. *Blood* **2004**, *103*, 4028–4035.
101. Sondermann, P.; Huber, R.; Oosthuizen, V.; Jacob, U. *Nature* **2000**, *406*, 267–273.
102. Radaev, S.; Motyka, S.; Fridman, W. H.; Sautes-Fridman, C.; Sun, P. D. *J. Biol. Chem.* **2001**, *276*, 16469–16477.
103. Ferrara, C.; Grau, S.; Jaeger, C.; Sondermann, P.; Bruenker, P.; Waldhauer, I.; Hennig, M.; Ruf, A.; Rufer, A. C.; Stihle, M.; Umana, P.; Benz, J. *Proc. Nat. Acad. Sci. U.S.A.* **2011**, *108*, 12669–12674.
104. Mizushima, T.; Yagi, H.; Takemoto, E.; Shibata-Koyama, M.; Isoda, Y.; Iida, S.; Masuda, K.; Satoh, M.; Kato, K. *Genes Cells* **2011**, *16*, 1071–1080.
105. Radaev, S.; Sun, P. *Mol. Immunol.* **2002**, *38*, 1073–1083.
106. Ramsland, P. A.; Farrugia, W.; Bradford, T. M.; Sardjono, C. T.; Esparon, S.; Trist, H. M.; Powell, M. S.; Tan, P. S.; Cendron, A. C.; Wines, B. D.; Scott, A. M.; Hogarth, P. M. *J. Immunol.* **2011**, *187*, 3208–3217.
107. Ward, E. S.; Ober, R. J., Multitasking by Exploitation of Intracellular Transport Functions: The Many Faces of Fcrn. In *Advances in Immunology*; Alt, F. W., Ed.; Academic Press: Waltham, MA, 2009; Vol. 103, pp 77–115.
108. Tesar, D. B.; Bjoerkman, P. J. *Curr. Opin. Struct. Biol.* **2010**, *20*, 226–233.
109. Kuo, T. T.; de Muinck, E. J.; Claypool, S. M.; Yoshida, M.; Nagaishi, T.; Aveson, V. G.; Lencer, W. I.; Blumberg, R. S. *J. Biol. Chem.* **2009**, *284*, 8292–8300.
110. Rath, T.; Kuo, T. T.; Baker, K.; Qiao, S. W.; Kobayashi, K.; Yoshida, M.; Roopenian, D.; Fiebiger, E.; Lencer, W. I.; Blumberg, R. S. *J. Clin. Immunol.* **2013**, *33* (Suppl 1), S9–17.
111. Mathiesen, L.; Nielsen, L. K.; Andersen, J. T.; Grevys, A.; Sandlie, I.; Michaelsen, T. E.; Hedegaard, M.; Knudsen, L. E.; Dziegiel, M. H. *Blood* **2013**, *122*, 1174–1181.
112. Einarsdottir, H. K.; Selman, M. H.; Kapur, R.; Scherjon, S.; Koeleman, C. A.; Deelder, A. M.; van der Schoot, C. E.; Vidarsson, G.; Wuhrer, M. *Glycoconjugate J.* **2013**, *30*, 147–157.
113. Lee, C. H.; Woo, J. H.; Cho, K. K.; Kang, S. H.; Kang, S. K.; Choi, Y. J. *Biotechnol. Appl. Biochem.* **2007**, *46*, 211–217.
114. Vllasaliu, D.; Alexander, C.; Garnett, M.; Eaton, M.; Stolnik, S. *J. Controlled Release* **2012**, *158*, 479–486.
115. Vallee, S.; Rakhe, S.; Reidy, T.; Walker, S.; Lu, Q.; Sakorafas, P.; Low, S.; Bitonti, A. *J. Interferon Cytokine Res.* **2012**, *32*, 178–184.
116. Stapleton, N. M.; Andersen, J. T.; Stemerding, A. M.; Bjarnarson, S. P.; Verheul, R. C.; Gerritsen, J.; Zhao, Y.; Kleijer, M.; Sandlie, I.; de Haas, M.; Jonsdottir, I.; van der Schoot, C. E.; Vidarsson, G. *Nat. Commun.* **2011**, *2*, 599.

117. Oganesyan, V.; Damschroder, M. M.; Woods, R. M.; Cook, K. E.; Wu, H.; Dall'Acqua, W. F. *Mol. Immunol.* **2009**, *46*, 1750–1755.
118. Zalevsky, J.; Chamberlain, A. K.; Horton, H. M.; Karki, S.; Leung, I. W. L.; Sproule, T. J.; Lazar, G. A.; Roopenian, D. C.; Desjarlais, J. R. *Nat. Biotechnol.* **2010**, *28*, 157–159.
119. Kapur, R.; Einarsdottir, H. K.; Vidarsson, G. *Immunol. Lett.* **2014**.
120. Jones, A. J. S.; Papac, D. I.; Chin, E. H.; Keck, R.; Baughman, S. A.; Lin, Y. S.; Kneer, J.; Battersby, J. E. *Glycobiology* **2007**, *17*, 529–540.
121. Keck, R.; Nayak, N.; Lerner, L.; Raju, S.; Ma, S.; Schreitmueller, T.; Chamow, S.; Moorhouse, K.; Kotts, C.; Jones, A. *Biologicals* **2008**, *36*, 49–60.
122. Zipfel, P. F. *Immunol. Lett.* **2009**, *126*, 1–7.
123. Degn, S. E.; Thiel, S. *Scand. J. Immunol.* **2013**, *78*, 181–193.
124. Voice, J. K.; Lachmann, P. J. *Eur. J. Immunol.* **1997**, *27*, 2514–2523.
125. Natsume, A.; In, M.; Takamura, H.; Nakagawa, T.; Shimizu, Y.; Kitajima, K.; Wakitani, M.; Ohta, S.; Satoh, M.; Shitara, K.; Niwa, R. *Cancer Res.* **2008**, *68*, 3863–3872.
126. Idusogie, E. E.; Wong, P. Y.; Presta, L. G.; Gazzano-Santoro, H.; Totpal, K.; Ultsch, M.; Mulkerrin, M. G. *J. Immunol.* **2001**, *166*, 2571–2575.
127. Natsume, A.; Shimizu-Yokoyama, Y.; Satoh, M.; Shitara, K.; Niwa, R. *Cancer Sci.* **2009**, *100*, 2411–2418.
128. Moore, G. L.; Chen, H.; Karki, S.; Lazar, G. A. *mAbs* **2010**, *2*, 181–189.
129. Aalberse, R. C.; Stapel, S. O.; Schuurman, J.; Rispens, T. *Clin. Exp. Allergy* **2009**, *39*, 469–477.
130. Morgan, A.; Jones, N. D.; Nesbitt, A. M.; Chaplin, L.; Bodmer, M. W.; Emtage, J. S. *Immunol.* **1995**, *86*, 319–324.
131. Diebolder, C. A.; Beurskens, F. J.; de Jong, R. N.; Koning, R. I.; Strumane, K.; Lindorfer, M. A.; Voorhorst, M.; Ugurlar, D.; Rosati, S.; Heck, A. J.; van de Winkel, J. G.; Wilson, I. A.; Koster, A. J.; Taylor, R. P.; Saphire, E. O.; Burton, D. R.; Schuurman, J.; Gros, P.; Parren, P. W. *Science* **2014**, *343*, 1260–1263.
132. Shields, R. L.; Lai, J.; Keck, R.; O'Connell, L. Y.; Hong, K.; Meng, Y. G.; Weikert, S. H. A.; Presta, L. G. *J. Biol. Chem.* **2002**, *277*, 26733–26740.
133. Davies, J.; Jiang, L. Y.; Pan, L. Z.; LaBarre, M. J.; Anderson, D.; Reff, M. *Biotechnol. Bioeng.* **2001**, *74*, 288–294.
134. Ferrara, C.; Brunker, P.; Suter, T.; Moser, S.; Puntener, U.; Umana, P. *Biotechnol. Bioeng.* **2006**, *93*, 851–861.
135. Peipp, M.; van Bueren, J. J. L.; Schneider-Merck, T.; Bleeker, W. W. K.; Dechant, M.; Beyer, T.; Repp, R.; van Berkel, P. H. C.; Vink, T.; van de Winkel, J. G. J.; Parren, P. W. H. I.; Valerius, T. *Blood* **2008**, *112*, 2390–2399.
136. Shinkawa, T.; Nakamura, K.; Yamane, N.; Shoji-Hosaka, E.; Kanda, Y.; Sakurada, M.; Uchida, K.; Anazawa, H.; Satoh, M.; Yamasaki, M.; Hanai, N.; Shitara, K. *J. Biol. Chem.* **2003**, *278*, 3466–3473.
137. Yamada, E.; Tsukamoto, Y.; Sasaki, R.; Yagyu, K.; Takahashi, N. *Glycoconjugate J.* **1997**, *14*, 401–405.

138. Ruhaak, L. R.; Uh, H.-W.; Beekman, M.; Hokke, C. H.; Westendorp, R. G. J.; Houwing-Duistermaat, J.; Wuhrer, M.; Deelder, A. M.; Sagboom, P. E. *J. Proteome Res.* **2011**, *10*, 1667–1674.
139. Holland, M.; Takada, K.; Okumoto, T.; Takahashi, N.; Kato, K.; Adu, D.; Ben-Smith, A.; Harper, L.; Savage, C. O. S.; Jefferis, R. *Clin. Exp. Immunol.* **2002**, *129*, 183–190.
140. Wuhrer, M.; Porcelijn, L.; Kapur, R.; Koeleman, C. A. M.; Deelder, A. M.; de Haas, M.; Vidarsson, G. *J. Proteome Res.* **2009**, *8*, 450–456.
141. Collins, E. S.; Galligan, M. C.; Saldova, R.; Adamczyk, B.; Abrahams, J. L.; Campbell, M. P.; Ng, C. T.; Veale, D. J.; Murphy, T. B.; Rudd, P. M.; Fitzgerald, O. *Rheumatology (Oxford, England)* **2013**, *52*, 1572–1582.
142. Farooq, M.; Takahashi, N.; Drayson, M.; Lund, J.; Jefferis, R. *Adv. Exp. Med. Biol.* **1998**, *435*, 95–103.
143. van Berkel, P. H. C.; Gerritsen, J.; Perdok, G.; Valbjorn, J.; Vink, T.; van de Winkel, J. G. J.; Parren, P. W. H. I. *Biotechnol. Prog.* **2009**, *25*, 244–251.
144. Wacker, C.; Berger, C. N.; Girard, P.; Meier, R. *Eur. J. Pharm. Biopharm.* **2011**, *79*, 503–507.
145. Raju, T. S.; Jordan, R. E. *mAbs* **2012**, *4*, 385–391.
146. Yamaguchi, Y.; Nishimura, M.; Nagano, M.; Yagi, H.; Sasakawa, H.; Uchida, K.; Shitara, K.; Kato, K. *Biochem. Biophys. Acta* **2006**, *1760*, 693–700.
147. Barb, A. W.; Meng, L.; Gao, Z.; Johnson, R. W.; Moremen, K. W.; Prestegard, J. H. *Biochemistry* **2012**, *51*, 4618–4626.
148. Houde, D.; Engen, J. R. *Methods. Mol. Biol.* **2013**, *988*, 269–289.
149. Lund, J.; Takahashi, N.; Pound, J. D.; Goodall, M.; Jefferis, R. *J. Immunol.* **1996**, *157*, 4963–4969.
150. Boyd, P. N.; Lines, A. C.; Patel, A. K. *Mol. Immunol.* **1995**, *32*, 1311–1318.
151. Sondermann, P.; Pincetic, A.; Maamary, J.; Lammens, K.; Ravetch, J. V. *Proc. Natl. Acad. Sci. U.S.A.* **2013**, *110*, 9868–9872.
152. Oaks, M.; Taylor, S.; Shaffer, J. *Oncoimmunology* **2013**, *2*, e24841.
153. Bayry, J.; Bansal, K.; Kazatchkine, M. D.; Kaveri, S. V. *Proc. Natl. Acad. Sci. U.S.A.* **2009**, *106*, E24–E24.
154. Kasermann, F.; Boerema, D. J.; Ruegsegger, M.; Hofmann, A.; Wymann, S.; Zuercher, A. W.; Miescher, S. *PLoS One* **2012**, *7*, e37243.
155. Yu, X.; Vasiljevic, S.; Mitchell, D. A.; Crispin, M.; Scanlan, C. N. *J. Mol. Biol.* **2013**, *425*, 1253–1258.
156. Stanfield, R. L. *Methods Mol. Biol.* **2014**, *1131*, 395–406.
157. Famm, K.; Hansen, L.; Christ, D.; Winter, G. *J Mol Biol* **2008**, *376*, 926–931.
158. Kim, S. J.; Hong, H. J. *Methods Mol. Biol.* **2012**, *907*, 247–257.
159. Dudgeon, K.; Rouet, R.; Famm, K.; Christ, D. *Methods Mol. Biol.* **2012**, *911*, 383–397.
160. Sagawa, T.; Oda, M.; Morii, H.; Takizawa, H.; Kozono, H.; Azuma, T. *Mol. Immunol.* **2005**, *42*, 9–18.
161. Sanguineti, S.; Centeno Crowley, J. M.; Lodeiro Merlo, M. F.; Cerutti, M. L.; Wilson, I. A.; Goldbaum, F. A.; Stanfield, R. L.; de Prat-Gay, G. *J. Mol. Biol.* **2007**, *370*, 183–195.
162. Keskin, O. *BMC Struct. Biol.* **2007**, *7*, 31.

163. Makabe, K.; Nakanishi, T.; Tsumoto, K.; Tanaka, Y.; Kondo, H.; Umetsu, M.; Sone, Y.; Asano, R.; Kumagai, I. *J. Biol. Chem.* **2008**, *283*, 1156–1166.
164. Asano, R.; Nakayama, M.; Kawaguchi, H.; Kubota, T.; Nakanishi, T.; Umetsu, M.; Hayashi, H.; Katayose, Y.; Unno, M.; Kudo, T.; Kumagai, I. *FEBS J.* **2012**, *279*, 223–233.
165. Stanfield, R. L.; Zemla, A.; Wilson, I. A.; Rupp, B. *J. Mol. Biol.* **2006**, *357*, 1566–1574.
166. Pritsch, O.; Magnac, C.; Dumas, G.; Bouvet, J. P.; Alzari, P.; Dighiero, G. *Eur. J. Immunol.* **2000**, *30*, 3387–3395.
167. Mompo, S. M.; Gonzalez-Fernandez, A. *Methods Mol. Biol.* **2014**, *1060*, 245–276.
168. Green, L. L. *Curr. Drug Discovery Technol.* **2014**, *11*, 74–84.
169. Woof, J. M.; Russell, M. W. *Mucosal Immunol.* **2011**, *4*, 590–597.
170. Bakema, J. E.; van Egmond, M. *mAbs* **2011**, *3*, 352–361.
171. Lohse, S.; Derer, S.; Beyer, T.; Klausz, K.; Peipp, M.; Leusen, J. H. W.; van de Winkel, J. G. J.; Dechant, M.; Valerius, T. *J. Immunol.* **2011**, *186*, 3770–3778.
172. Gautam, S.; Loh, K.-C. *Biotechnol. Adv.* **2011**, *29*, 840–849.
173. Tchoudakova, A.; Hensel, F.; Murillo, A.; Eng, B.; Foley, M.; Smith, L.; Schoenen, F.; Hildebrand, A.; Kelter, A.-R.; Ilag, L. L.; Vollmers, H. P.; Brandlein, S.; McIninch, J.; Chon, J.; Lee, G.; Cacciuttolo, M. *mAbs* **2009**, *1*, 163–171.
174. Gould, H. J.; Sutton, B. J. *Nat. Rev. Immunol.* **2008**, *8*, 205–217.
175. Odani-Kawabata, N.; Takai-Imamura, M.; Katsuta, O.; Nakamura, H.; Nishioka, K.; Funahashi, K.; Matsubara, T.; Sasano, M.; Aono, H. *BMC Musculoskeletal Disord.* **2010**, *11*, 221.
176. Spillner, E.; Plum, M.; Blank, S.; Miehe, M.; Singer, J.; Braren, I. *Cancer Immunol. Immunother.* **2012**, *61*, 1565–1573.
177. Rudman, S. M.; Josephs, D. H.; Cambrook, H.; Karagiannis, P.; Gilbert, A. E.; Dodev, T.; Hunt, J.; Koers, A.; Montes, A.; Taams, L.; Canevari, S.; Figini, M.; Blower, P. J.; Beavil, A. J.; Nicodemus, C. F.; Corrigan, C.; Kaye, S. B.; Nestle, F. O.; Gould, H. J.; Spicer, J. F.; Karagiannis, S. N. *Clin. Exp. Allergy* **2011**, *41*, 1400–1413.
178. Karagiannis, S. N.; Josephs, D. H.; Karagiannis, P.; Gilbert, A. E.; Saul, L.; Rudman, S. M.; Dodev, T.; Koers, A.; Blower, P. J.; Corrigan, C.; Beavil, A. J.; Spicer, J. F.; Nestle, F. O.; Gould, H. J. *Cancer Immunol. Immunother.* **2012**, *61*, 1547–1564.
179. Desjarlais, J. R.; Lazar, G. A. *Exp. Cell Res.* **2011**, *317*, 1278–1285.

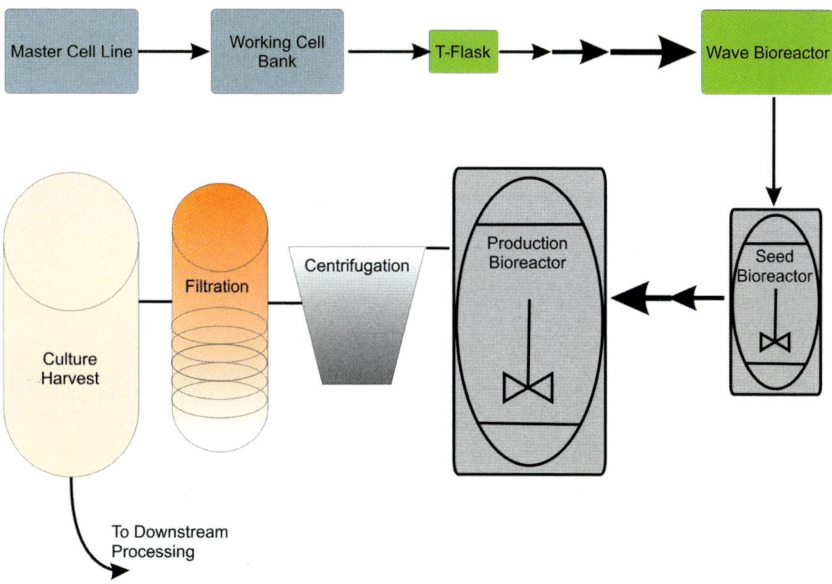

Figure 1. Representative upstream processing steps that may be used for monoclonal antibody production. (Chapter 1)

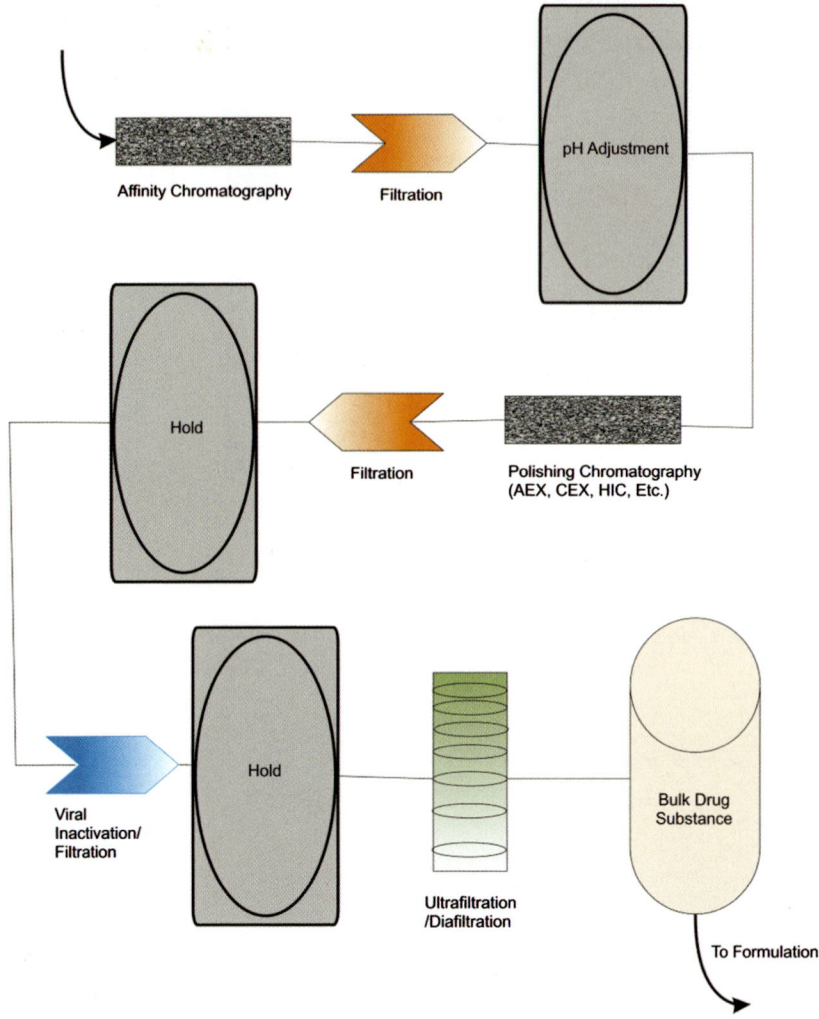

Figure 2. Representative downstream processing steps that may be used for monoclonal antibody production. Potential polishing chromatography steps include anion exchange chromatography (AEX), cation exchange chromatography (CEX), and hydrophobic interaction chromatography (HIC). (Chapter 1)

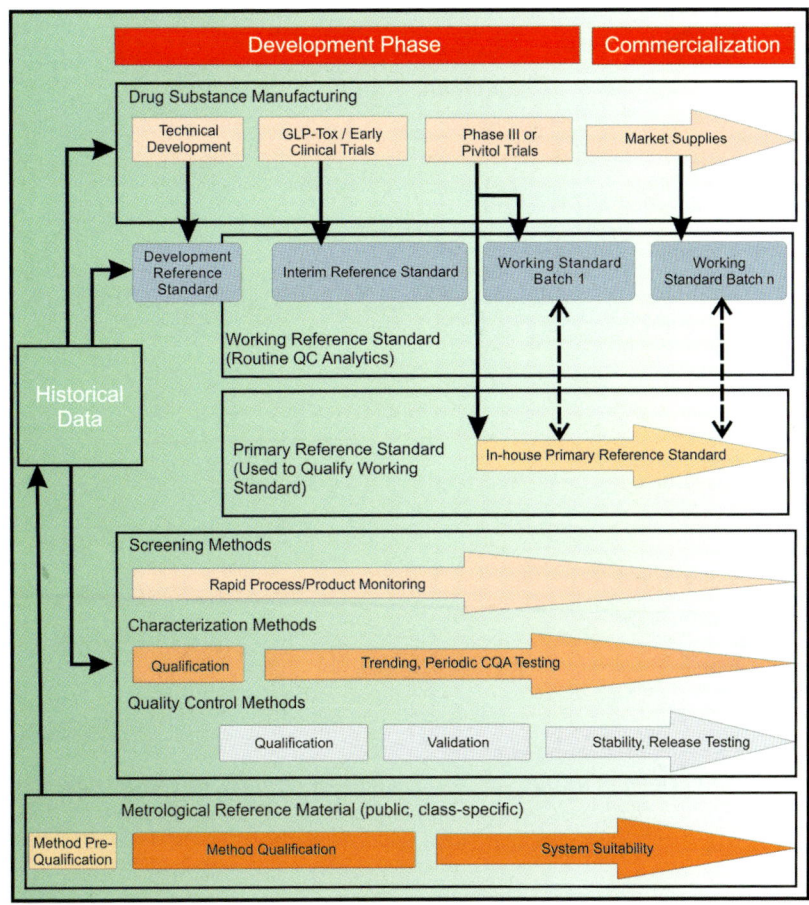

Figure 3. Representative monoclonal antibody lifecycle incorporating potential timelines for analytical method development, in-house reference standards, and potential supplementation with a metrological reference material. (Chapter 1)

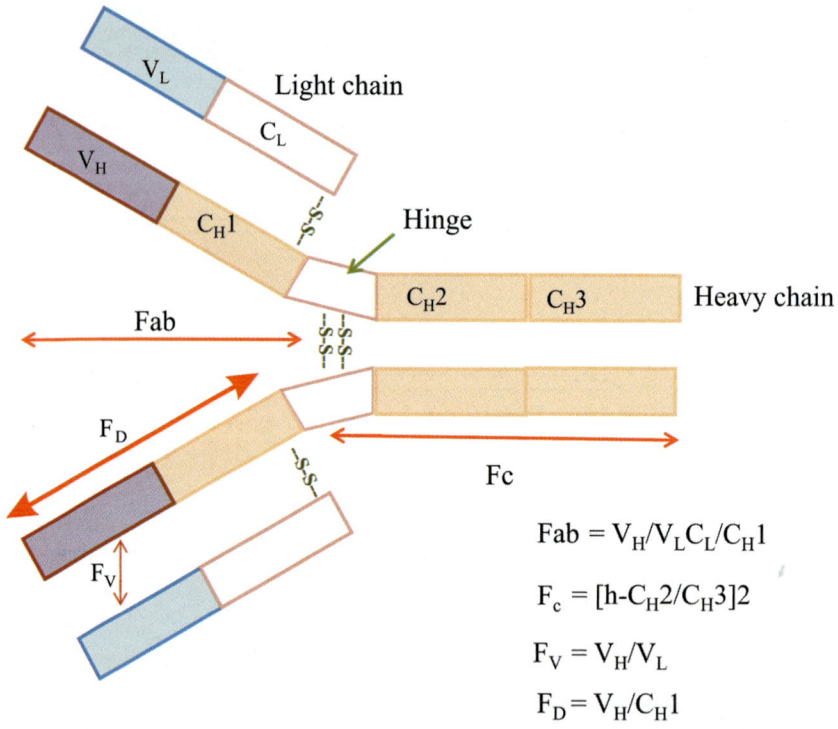

Figure 1. A cartoon of the four chain structure for an IgG1 molecule; inter-chain disulphide bridges only are shown. (Chapter 2)

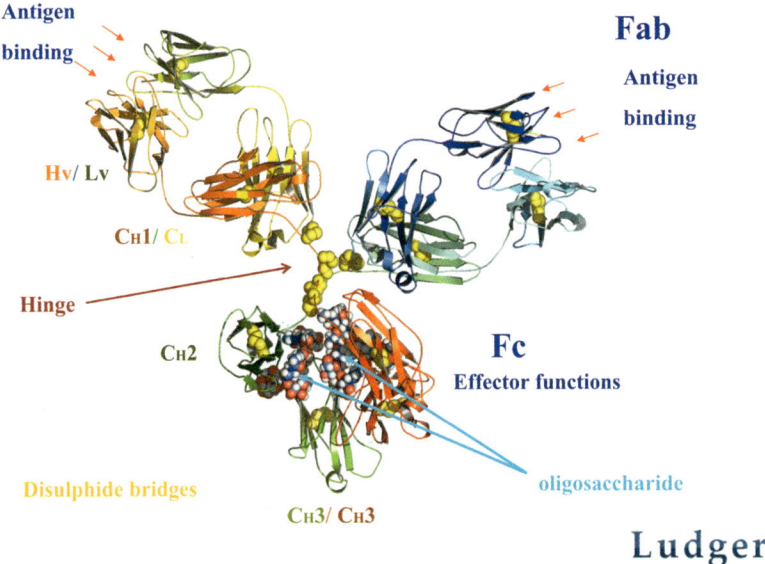

Figure 2. The domain structure of the IgG molecule. (Chapter 2)

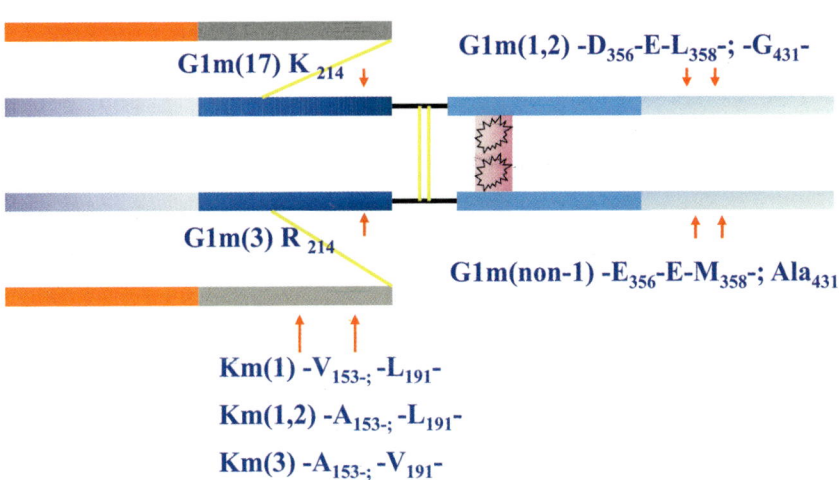

Figure 3. Sequence correlates of IgG1.G1m and Km allotypes. (Chapter 2)

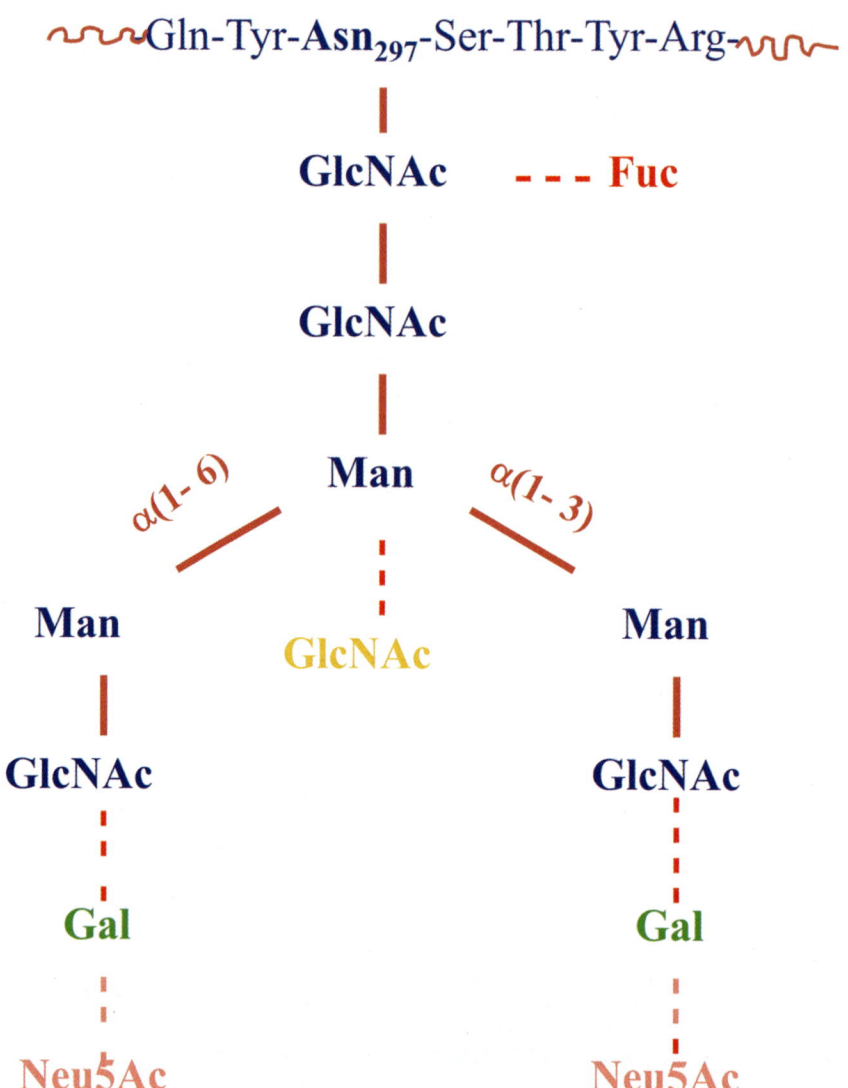

Figure 4. Representative IgG complex diantennnary oligosaccharides; comprised of a "core" -GlcNAc)2Man3(GlcNAc)2 heptasaccharide. (Chapter 2)

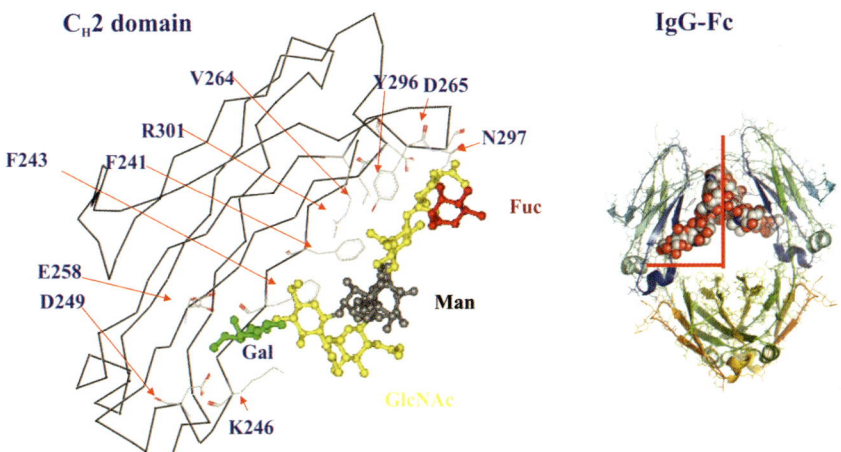

Figure 5. The IgG C_H2 domain showing amino acid residues contributing non-covalent interactions with the oligosaccharide. (Chapter 2)

Chapter 3

Heterogeneity of IgGs: Role of Production, Processing, and Storage on Structure and Function

Chris Barton,[1] David Spencer,[1] Sophia Levitskaya,[1] Jinhua Feng,[1] Reed Harris,[2] and Mark A. Schenerman*,[1]

[1]Analytical Biotechnology, MedImmune,
Gaithersburg, Maryland 20878, United States
[2]Genentech, Inc., South San Francisco, California 94080, United States
*E-mail: schenermanm@medimmune.com

Detailed monoclonal antibody (mAb) characterization tools have enabled the discovery of structural variations, including many that compromise functionality or have other undesired properties. Size, charge, glycosylation, and disulfide bonding variants; oxidized amino acid residues; and polypeptide chain truncations, extensions, and cleavage points have been identified. Product quality attributes, including detection techniques, published knowledge about the process origins, and quality impacts of these variants are summarized.

Introduction

Use in Diagnostics, Reagents, and Medicinals

Since the discovery of monoclonal antibodies (mAbs), the enormous value of this class of proteins has yet to be fully realized (*1*). First used as highly specific laboratory reagents, mAbs now support the *in vitro* diagnostics, imaging, and radiotherapy markets and have become therapeutic agents used in treatments for cancer, inflammatory and autoimmune diseases, infectious diseases, and cardiovascular indications. Datamonitor (*2*) estimates the molecular diagnostics market at $3 billion in 2010, and diagnostics sales are projected to grow at a 10 percent compound annual growth rate (CAGR) from 2010 to 2016. The estimated global sales of therapeutic antibodies, according to Evaluate Pharma (*3*), are $55 billion in 2012. They are forecast to grow at a 9 percent CAGR from 2012 to 2018 to $92 billion. mAbs are expected to have the fastest rate of growth (+8% CAGR) among all of the classes of branded pharmaceuticals between now and 2016.

mAbs are also extremely diverse proteins that are subject to considerable heterogeneity due to production, processing, and storage. This chapter will review the affect that upstream, downstream, formulation, and fill/finish process operations, as well as long-term storage, can have on structure and function.

The Need for Analytical Information

Our patients, as well as the federal health authorities, require that we produce a consistently safe and active drug product. Although the general properties of recombinant proteins are primarily determined by their molecular design, we continue to find that recombinant proteins, including recombinant antibodies, are highly heterogeneous, and that many of the variants have properties that can affect safety, immunogenicity, pharmacokinetics/pharmacodynamics, and/or bioactivity. The development functions need to establish a product's overall covalent structure, identify sources of variability, assess the likely patient impacts for each variable attribute, and establish a suitable control strategy, including process design and testing controls. Useful methods for identifying sources of variability, and their impacts and origins, are described in this chapter.

Demonstrating Product Quality

Recombinant mAb testing generally falls into three categories: quality control (QC) testing (in-process, release, and stability), extended characterization using research and development (R&D) methods, and high-throughput screening assays. Expectations for QC testing are described in guidance documents such as ICH Q6B, and defined by the Code of Federal Regulations (CFR). These expectations include tests for purity, potency, identity, strength, and general properties such as appearance, excipients, stabilizers, and pH. For the protein component, purity is usually a relative term, reported as a percentage of the total species detected. Potency may be expressed as a relative value, comparing the test article to a qualified in-house reference standard. Identity tests need to differentiate the product from others that may be produced or stored in the same facility. Strength may be reported as a concentration, usually based on an extinction coefficient. General tests are often based on compendia. QC tests are performed in a qualified good manufacturing practices (GMPs) laboratory using trained staff.

In-process QC testing is generally performed on process step samples to inform purity, such as by testing for endotoxin or bioburden (due to microbial contamination); monitor step yield; or assess removal of process-related impurities such as host cell proteins or purification column leachates. In-process tests are performed across a range of concentrations and buffer excipients, so multiproduct methods are generally needed. In-process tests can also be used instead of final batch tests for assessing the addition of excipients. Action limits are often useful for in-process tests, where a value that exceeds the limit requires a GMP investigation but does not automatically cause batch rejection.

Batch release testing is performed on samples from the finished drug substance and drug product. Testing should be performed on samples closest to the step that affects the product's characteristics. For example, glycosylation testing is appropriate for the drug substance step but does not need to be repeated for the drug product. Tests for visible particles are appropriate for drug products but not for the drug substance. Many tests, such as size- or charge-based methods, may appear on both drug substance and drug products because of the potential for drug substance and product processing to each affect those characteristics.

Release tests have acceptance criteria that define a quantitative range or qualitative profile assessment to which the batch test results must conform to enable release. Many release tests are also used to assess stability, particularly when they enable monitoring of degradation pathways such as aggregation and fragmentation, deamidation, or loss of potency.

Extended characterization methods use advanced tools to assess primary and higher order structure, as well as provide information regarding covalent modifications such as post-translational and degradative modifications. The extended methods may be needed to augment potency assays to confirm the proposed biological mechanism(s) of action (MoAs). Early in clinical development, it may be acceptable to include tests in the extended characterization panel that are hard to validate or require newer technologies, with the understanding that some of these methods may be added to the QC tests later in product development, such as when the final to-be-commercialized control strategy is established.

The combined QC and extended characterization methods form the nucleus of any comparability assessment, where the sponsor is asked to confirm that a process or production facility change did not adversely affect the safety or efficacy compared to earlier process materials. These extended characterization methods are often used to determine the critical quality attributes (CQAs), and many of these methods are adapted for use in QC release testing and for stability assessments if the "routine" assays are unable to resolve and quantitate the CQA. A different sort of method adaptation occurs when key assays are converted to a high-throughput mode, such as the use of automated capillary electrophoresis (CE)-based or chip-based techniques that provide lower resolution but remain useful for process design.

The term "quality" remains a challenge to define, but most sponsors apply the guidelines in the International Conference on Harmonisation (ICH) guidance documents and CFR. Beyond health authority requirements, responsible sponsors seek to understand which attributes need to be controlled by the process or confirmed by batch testing to demonstrate that they are providing a safe and efficacious product over the entire shelf life. Early in clinical development, the quality emphasis is on safety, bioactivity, and preservation of the analytical profile relative to an intended profile that may be determined by testing materials used in toxicology and other preclinical studies. In early development, platform tests are often used with minimal molecule-specific re-optimization. As development proceeds, tests are refined based on additional method development opportunities;

potency assays are further optimized to reflect MoA, often using cell-based assays; and the structural and functional characteristics of the minor component are identified. The combined product knowledge enables assignment of the CQAs. Final QC and extended characterization tests usually are established by the time a sponsor begins pivotal (Phase 3) clinical studies, and are designed to ensure proper control of CQAs, particularly for high-impact CQAs or those that lack suitable process controls (see Figure 1 for details). Additional general methods are required by regulations.

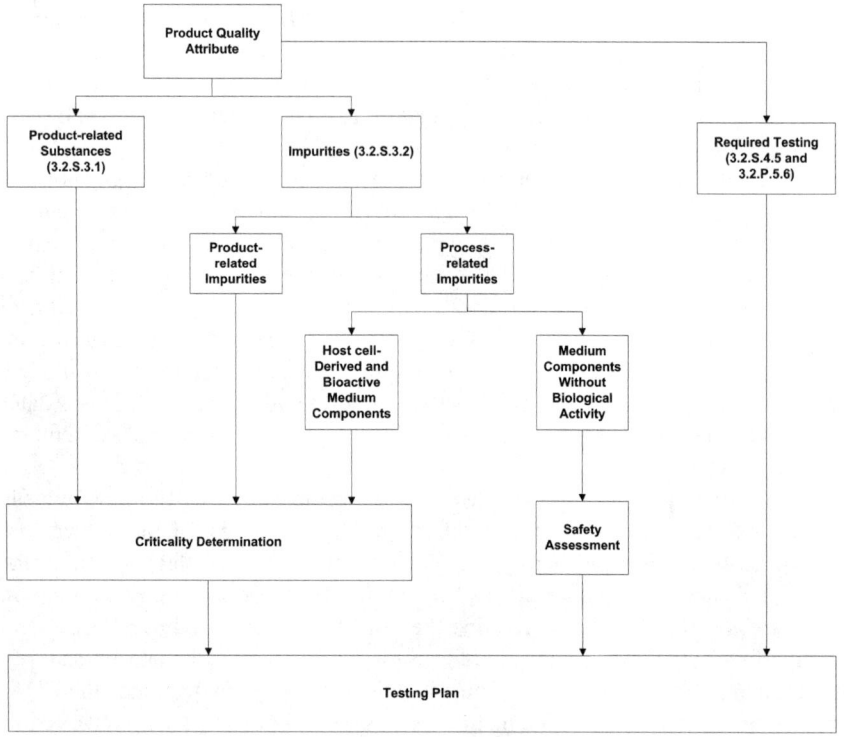

Figure 1. Pathway for development of a testing plan based on product quality attributes. Reproduced with permission from Schenerman et al. (139). Copyright © 2009 by John Wiley & Sons, Inc.

Identification of Critical Quality Attributes

A CQA is defined in ICH Q2(R1) as a "physical, chemical, biological, or microbiological property or characteristic that should be within an appropriate limit, range, or distribution to ensure the desired product quality." One approach to assess if a molecular variant or impurity is deemed "critical" was developed at Genentech and incorporated into the A-Mab case study (*4*). This approach examines the impact of that attribute on:

- Biological activity using one or more MoA-relevant methods.
- Pharmacokinetics, usually either the area under a concentration-time curve from *in vivo* studies or modeled using FcRn assays.
- Relative immunogenicity risk, which often requires subject matter judgment based on *in vivo* and/or *in silico* models, relevant literature, and/or clinical experience.
- Safety impacts, which are usually based on rare but high-impact clinical events involving related molecules, or from nonclinical models.

The attributes are assessed on a quantitative scale from low to very high impact. The highest impact value is combined with an uncertainty factor to determine the overall CQA severity score, with high severity attributes generally requiring batch testing, moderate severity attributes often being subjected to batch testing if the process capability or stability is poor relative to an acceptable range, and low severity attributes often being part of comparability assessments or monitoring programs or not being tested further. A list of CQAs for a mAb is shown in Table 1. Further details on CQAs assessment and control strategies for IgGs are presented in the QbD chapter/Volume 1, Chapter 5.

Technical Challenges

mAbs present challenges for molecular characterization due to their overall size, with often approximately 1,450 amino acid residues found in two identical light chains and two identical heavy chains, as well as their molecular heterogeneity. However, common tools such as mass spectrometry or peptide mapping can differentiate molecular variants or modified forms. For example, isomerization of single aspartyl residue in trastuzumab causes complete resolution of the modified form from the main form in a cation-exchange assay (*5*). However, assignment of this modification was challenging because even the modified fraction retains the expected structure for half of the material. Some modifications co-migrate with other unrelated forms, again diminishing the differentiation from the main form. Single modification sites often can be identified at low percentage levels, particularly when compared to an unmodified form, using peptide map approaches. A different challenge exists when a single type of modification is found at low levels across a large number of sites, such as glycation; modifications of this type are often detected only in the intact form using mass spectrometry, and assignment of specific sites remains a laborious task. Despite these challenges, several review articles (*6–8*) have summarized the types of modifications found on recombinant antibodies, including deamidation and isomerization, unpaired cysteine (Cys) residues, and chain terminations or extensions, as described below.

Table 1. Product Attribute Categories and Tests

Category	Product Attribute	Test	Purpose	ICH Q6B Category	Process Operations Affecting Critical Quality Attributes
Size	Aggregation	Size exclusion chromatography (SEC)	Detect product-related impurities (fragments, aggregates)	Identity, Purity, Stability	Upstream, downstream, formulation, fill/finish, and storage
		Gel electrophoresis	Detect product-related impurities (fragments, aggregates)	Identity, Purity, Stability	Upstream, downstream, formulation, fill/finish, and storage
		Analytical ultracentrifugation	Detect product-related impurities (fragments, aggregates)	Identity, Purity	Upstream, downstream, formulation, fill/finish, and storage
		SEC with multi-angle light scattering (MALS)	Detect product-related impurities (fragments, aggregates)	Identity, Purity	Upstream, downstream, formulation, fill/finish, and storage
	Truncation	Ion-exchange chromatography (with and without carboxypeptidase)	Detect charge isoforms	Identity, Purity, Stability	Upstream and storage
		Isoelectric focusing (with and without carboxypeptidase)	Assess pattern of charge isoforms	Identity, Purity, Stability	Upstream and storage
		Peptide mapping with mass spectrometry	Verify primary structure and identify post-translational modifications	Identity, Purity, Stability	Upstream and storage

Category	Product Attribute	Test	Purpose	ICH Q6B Category	Process Operations Affecting Critical Quality Attributes
	Fragmentation	SEC	Detect product-related impurities (fragments, aggregates)	Identity, Purity, Stability	Upstream, downstream, and storage
		Gel electrophoresis	Detect product-related impurities (fragments, aggregates)	Identity, Purity, Stability	Upstream, downstream, and storage
		Analytical ultracentrifugation	Detect product-related aggregates and fragments	Identity, Purity	Upstream, downstream, and storage
		Peptide mapping with mass spectrometry	Verify primary structure and identify post-translational modifications	Identity, Purity, Stability	Upstream, downstream, and storage
		Reversed phase high-performance liquid chromatography (HPLC)	Detect product-related fragmentation and isomerization	Identity, Purity	Upstream, downstream, and storage
	Total mass	Quadrupole-time-of-flight (Q-TOF) mass spectrometry	Measure total molecule mass	Identity	Upstream
		Light scattering	Measure total molecule mass	Identity	Upstream

Continued on next page.

Table 1. (Continued). Product Attribute Categories and Tests

Category	Product Attribute	Test	Purpose	ICH Q6B Category	Process Operations Affecting Critical Quality Attributes
Charge	Deamidation	Ion-exchange chromatography	Detect charge isoforms	Identity, Purity, Stability	Upstream, downstream, and storage
		Isoelectric focusing	Assess pattern of charge isoforms	Identity, Purity, Stability	Upstream, downstream, and storage
		Peptide mapping with mass spectrometry	Verify primary structure and identify post-translational modifications	Identity, Purity, Stability	Upstream, downstream, and storage
	Sialylation	Sialic acid content	Measure sialic acid content	Identity	Upstream
Primary structure	Identity	Peptide mapping with mass spectrometry	Verify primary structure and identify post-translational modifications	Identity, Purity, Stability	Upstream and storage
		Amino acid analysis	Measure amino acid composition	Identity	Upstream
		N-terminal Sequencing	Assess N-terminal sequence	Identity	Upstream
		Immunoassay	Show specific immunoreactivity	Identity	Upstream

Category	Product Attribute	Test	Purpose	ICH Q6B Category	Process Operations Affecting Critical Quality Attributes
Post-translational modifications	Disulfide bonds	Disulfide bond determination	Verify correct disulfide bond locations	Identity	Upstream and storage
	Glycation	Peptide mapping with mass spectrometry	Verify primary structure and identify post-translational modifications	Identity, Purity, Stability	Upstream and storage
		Boronate HPLC	Detect glycation	Identity, Purity	Upstream and storage
	Glycosylation	Monosaccharide composition analysis	Quantify monosaccharide composition	Identity	Upstream
		Oligosaccharide profile	Assess pattern of oligosaccharide profile	Identity	Upstream
		Galactose content	Measure galactose content	Identity	Upstream
		Fucose content	Measure fucose content	Identity	Upstream
	Isomerization	Reversed phase HPLC	Detect product-related fragmentation and isomerization	Identity, Purity	Upstream, downstream, and storage

Continued on next page.

Table 1. (Continued). Product Attribute Categories and Tests

Category	Product Attribute	Test	Purpose	ICH Q6B Category	Process Operations Affecting Critical Quality Attributes
	Oxidation	Peptide mapping with mass spectrometry	Verify primary structure and identify post-translational modifications	Identity, Purity, Impurity	Upstream, downstream, and storage
	Thioether link	Gel electrophoresis	Detect product-related impurities (fragments, aggregates)	Identity, Purity, Stability	Upstream, downstream, and storage
Higher order structure	Conformation	Circular dichroism	Detect secondary structure changes	Identity	Upstream
		Fourier transform infrared spectroscopy	Detect secondary/tertiary structure changes	Identity	Upstream
		Differential scanning calorimetry	Detect tertiary structure changes	Identity	Upstream
		Bioassay/binding assay	Measure bioactivity	Potency, Identity	Upstream, downstream, formulation, fill/finish, and storage
		X-ray crystallography	Detect tertiary structure changes	Identity	Upstream
	Effector function	Fc receptor binding	Determine Fc function	Potency	Upstream
		Complement binding	Determine Fc function	Potency	Upstream

Product Attribute Assessment

Size

Aggregation

Under ideal conditions, an antibody would be a biochemically active single molecule (monomer) folded in a unique native structure. In reality, subtle differences in charge, structure, conformation, and morphology can promote protein-protein interactions (PPIs) and cause formation of relatively stable groups of antibody molecules that are collectively termed aggregates (discussed in detail in the Aggregation chapter/Volume 3, Chapter 5). In the extreme case of aggregate formation, covalent bonds through disulfide bridges or oxidized tyrosine are formed. Protein aggregation may be a challenge throughout upstream, downstream, formulation, and drug delivery development process operations, as well as during long-term storage, due to the possibility of biochemical modification or mechanical stress that lead to PPIs (*9–12*).

Several recent publications review factors affecting aggregation and propose a classification of the possible mechanisms (*13–19*). Aggregation formation via an unfolded protein state is summarized by the classical Lumry-Eyring equation: $N \Leftrightarrow U \rightarrow A$, where N is native, U is unfolded, and A is aggregated protein (*20*). This has been elaborated in recent works to reflect a complicated, multifactorial process of protein aggregation. The aggregation mechanisms include at least two steps with different dynamics—initial nucleation and secondary clusterization (*18*). Soluble, reversible aggregates are formed through weak PPIs, such as electrostatic, hydrophobic, and van der Waal's interactions (*10, 14, 21–23*). Irreversible aggregates are formed from structurally altered species of IgG, represented by unfolded or partially unfolded covalently bound monomers and fragments (*24*). Irreversible aggregates have the potential to serve as nuclei for accumulation of large multimers and insoluble particles and can be associated with long-term aggregation.

The most popular analytical tool for detecting aggregation in antibody products is size exclusion chromatography (SEC) (*25, 26*). SEC is a relatively simple method that separates molecules by their hydrodynamic volumes, with larger molecules (unordered multimers, tetramers, trimers and dimers) eluting first, followed by monomer and then smaller fragments. As the different species elute, they are detected and quantitated by UV absorbance. For more information, light scattering detectors such as MALS (multi-angle light scattering), RALS (right-angle light scattering), and LALS (low-angle light scattering) can be added to SEC for determination of molecular weight (MW) of the components. Analytical ultracentrifugation (AUC), field flow fractionation (FFF), dynamic and static light scattering (DLS and SLS), and electrophoretic sodium dodecylsulfate (SDS) separation in gels are other conventional methods for aggregate assessment (*16, 27–29*). These methods can be used as complementary or orthogonal to SEC. Various spectroscopic (circular dichroism [CD], Fourier transform infrared spectroscopy [FTIR], Raman), mass spectrometry (liquid chromatography-mass spectrometry [LC-MS], Hydrogen/Deuterium Exchange [H/D], electrospray ionization [ESI]-time-of-flight mass spectrometry [TOF-MS]), microscopy

(Transmission Electron Microscopy [TEM], Atomic Force Microscopy [AFM]), small-angle neutron or X-ray scattering (SANS, SAXS), and nuclear magnetic resonance (NMR) methods have been applied for detailed structural characterization and to study mechanisms of aggregation (*30–35*).

Understanding protein aggregation in biopharmaceutical products is important due to the potential impact on purity, efficacy, and immunogenicity (*36–40*). Furthermore, the formation of aggregates on stability, especially in liquid solutions, may limit a product's shelf life. Control of aggregation and, when possible, elimination of factors favorable for aggregation are therefore objectives for antibody developers. Because the stability of antibody molecules is primarily determined by structural features, identification of aggregation-prone motifs in IgG sequences and molecular modeling are implemented as screening tools in the development of stable products (*41, 42*). Additionally, use of appropriate formulation excipients and process buffers is essential to mitigate aggregation (*43, 44*).

Truncation and Extensions

Truncation is a process in which amino acids are cleaved from either the C- or N-terminus of intact proteins. The most commonly observed truncation of mAbs is C-terminal lysine (Lys) truncation, also called "lysine clipping" or "lysine processing." Lys is expected to be the C-terminal residue of mAb heavy chains, but is often absent in mAbs purified from mammalian cell culture, such as the Chinese hamster ovary (CHO) cell lines. This discrepancy is due to the cleavage of the C-terminal Lys by enzymes (carboxypeptidases) in cell culture (*45*). The precise mechanism of the C-terminal Lys truncation is unknown, but several studies have shown that C-terminal Lys truncation occurs during mAb production in culture, both during protein processing in the endo-membrane system and following protein secretion into the cell culture medium (*46*).

Truncations other than C-terminal Lys truncation have been observed in mAbs. C-terminal α-amidation is a recently reported C-terminal modification involving the proline-glycine (Pro-Gly) terminus that remains after Lys processing. The terminus is susceptible to further processing of the Gly residue to produce a C-terminal Pro amide. (*47, 48*). This C-terminal Pro amidation is less common due to the initial requirement of the processing of both Lys and Gly on C-terminus. C-terminal arginine truncation was also reported for recombinant human erythropoietin (*49*) and two-chain tissue plasminogen activator (*45*).

C-terminal Lys truncation can be analyzed by a range of analytical techniques. This truncation mechanism results in variants containing either zero, one, or two C-terminal Lys residues on two mAb heavy chains. Since Lys is positively charged at physiological pH, the net charge of these C-terminal Lys variants increases in the order of 0 < 1 < 2 Lys. This leads to charge heterogeneity that can be resolved by charge separation techniques, such as ion-exchange chromatography (IEX) and isoelectric focusing (IEF). Both IEX and IEF (including gel and capillary isoelectric focusing) are widely used to detect and quantify the C-terminal Lys variants of therapeutic mAbs. C-terminal Lys truncation is not the sole cause for the charge heterogeneity for mAbs, with other post-translational modifications such as sialylation, N-terminal cyclization, C-terminal amidation, deamidation, and fragmentation described in other sections.

The removal of one C-terminal Lys residue decreases the MW by 128 Da. Therefore, the identification of C-terminal Lys variants can also be confirmed by MW determination using ESI (*50*) and matrix-assisted laser desorption ionization (MALDI) (*51*) mass spectrometry (*52*). To confirm the Lys truncation occurs at the C-terminus, peptide mapping combined with reversed-phase high-performance liquid chromatography and mass spectrometry (RP-HPLC-MS) is used.

The degree of C-terminal Lys truncation varies significantly depending on the fermentation process, including cell line (*53*), cell culture medium (*52*), trace elements such as copper and zinc (*46*), temperature, and duration (*46*). It has been reported that C-terminal Lys variants of a mAb all exhibit the same biological activity, consistent with truncation as a modification at a site distal from the antigen-binding and effector function domains (*52*). However, the heterogeneity of the C-terminal Lys variants is a sensitive indication for the manufacturing process change and should be monitored by the methods described above for product consistency.

N-terminal sequence extensions due to mis-cleavages within N-terminal leader sequences have also been reported. For example, the valine-histidine-serine (Ser) extension (*54*) adds a basic residue that enables resolution using charge-based methods. The additional basic residues do not affect potency or pharmacokinetics (*54*). Unspecified basic N-terminal extensions of one to eight residues have also been reported (*55*).

Fragmentation

Fragmentation is a degradative process in which a covalent bond in a protein is disrupted. The cleavage of a covalent bond can be categorized into two subclasses: peptide backbone cleavage or side chain cleavage (e.g., disulfide bond cleavage). This section covers peptide backbone fragmentation; disulfide bond fragmentation is covered later in the chapter.

Fragmentation of the peptide backbone may occur by hydrolysis through either a chemical or enzymatic reaction. The peptide backbone is extremely stable to non-enzymatic fragmentation under physiological conditions (neutral pH 7.4 and below 36°C). However, certain sites may become prone to fragmentation when the environment is favorable, based on the specific amino acid sequence, flexibility of the local peptide structure, buffer composition, pH, temperature, and the presence of metals or radicals. It is well known that peptide bonds in the hinge region of mAbs are susceptible to hydrolysis, generating antigen-binding fragments (Fabs) and Fab-Fc fragments (*56–60*). The fragmentation in the hinge region is accelerated in acidic or basic conditions (*59*), which implies an acid- or base-catalyzed mechanism leading to hinge region peptide bond cleavage. The fragmentation rate in the hinge region can be significantly increased in the presence of metal ions such as Cu^{2+}, Fe^{2+}, and Fe^{3+} (*61, 62*).

Non-enzymatic fragmentation outside of the hinge region is largely sequence-specific. Most of the peptide backbone fragmentation events occur at one of the following residues: aspartic acid (Asp), Gly, Ser, threonine (Thr), Cys, or asparagine (Asn). It has been hypothesized that the side chains of these residues (with the exception of Gly) can facilitate peptide bond cleavage via specific mechanisms and that Asp-Pro is a particularly labile motif (63). However, primary structure is not the only determinant in mAb fragmentation. The secondary and higher order environment around the cleavage site is also important. Accordingly, exposed peptide bonds in domain-domain interfaces, either at the edges of β-sheets or in the loops that connect the β-sheets, are reported to be susceptible to fragmentation (64). These regions include the solvent-exposed, flexible loops of mAb complementarity-determining regions (CDRs). It is also worth noting that IgG1 is more susceptible to fragmentation, compared with IgG2 and IgG4 under pH 5 to pH 5.5 conditions (65), suggesting a conformational role in fragmentation rates.

Fragmentation by proteolytic activity is also possible during mAb production because of the assortment of proteolytic enzymes that occur among the host cell proteins present in the harvested cell culture fluid (66). During purification of therapeutic mAbs, the level of host cell proteins is expected to be reduced to acceptable levels by removing proteolytic enzymes. However, proteolytic fragmentation can occur during downstream processing prior to protease removal, and it has been reported that fragmentation due to residual proteolytic activity can occur in highly purified mAbs (67).

Cleavage of a peptide bond can significantly alter the properties of the molecule and can be detected by various analytical methods based on size, hydrophobicity, and charge. The most common methods used to monitor fragmentation are size-based methods such as SEC, SDS-polyacrylamide gel electrophoresis (SDS-PAGE), and capillary sodium dodecylsulfate electrophoresis (cSDS), or methods that exploit differing hydrophobicity such as RP-HPLC (Separation chapter/Volume 2, Chapter 5).

SEC is a critical method to monitor the aggregates in therapeutic mAbs, and it also provides information about fragmentation, such as the Fab and Fab-Fc fragments resulting from hinge region cleavage. However, as a non-denaturing technique, SEC is unable to detect peptide bond cleavages that generate fragments held together by other covalent bonds (e.g., disulfide bonds) or non-covalent interactions.

SDS-PAGE or its capillary counterpart, cSDS, provides excellent resolution of fragments, and these methods are widely used to monitor overall fragmentation in mAbs. Both methods are run under denaturing (and potentially reducing) conditions, enabling identification of fragments not observed by SEC. cSDS is now commonly used in the pharmaceutical industry due to the straightforward quantification and improved resolution compared with the traditional slab gel SDS-PAGE.

RP-HPLC is an important analytical method for separating fragments based on hydrophobicity differences, and it is even more powerful when connected to ESI mass spectrometry for the direct identification of the cleavage sites. However, quantitation can be hindered by low resolution when compared with cSDS.

Fragmentation, depending on the site, may affect the bioactivity, pharmacokinetic (PK) properties, and stability of a protein. Fragmentation in the mAb CDRs may affect target binding. Fragmentation in the hinge region may have implications on the function of a mAb molecule: the Fab fragment will be devoid of any Fc-mediated effector function and have a reduced circulation half-time; the Fc-Fab fragment may not be potent if interaction with the target receptor requires both Fab arms. Similarly, fragmentation in the constant regions of mAbs may have an effect on either the Fc-mediated effector function or on the circulation half-time. The relationship between cleavage and potency can be unclear. For these reasons, the effect of fragmentation on the function of a mAb has to be evaluated on a case-by-case basis, depending on whether cleavage sites are observed in the variable or constant regions, and on the MoA of the molecule.

Fragmentation affects the mAb purity and can indicate the degradation of the molecule during long-term storage. To ensure that the potency and purity of therapeutic mAb products is maintained, fragmentation should be minimized and controlled below a certain level. To facilitate this, mAb products are typically formulated in neutral pH 6.0 buffers containing a low level of metal ions, and stored in frozen or refrigerated conditions.

Total Mass

The total mass (MW) of the mAb is a key indicator of protein size. As discussed above, aggregation, fragmentation, and truncation all affect the mAb total mass, as do post-translational modifications such as glycosylation. The presence and levels of these size variants may be influenced by upstream and downstream processes, as well as by long-term storage, and therefore require monitoring. In addition to the techniques described above to measure aggregation, fragmentation, and truncation, there are a number of dedicated tools available to measure mAb total mass. Mass spectrometry techniques including ESI-MS (*50*) and MALDI-TOF-MS (*51*) can measure the mass of a mAb to within a few Daltons, depending on the instrument. ESI-MS is a particularly versatile technique that can measure the mass of the mAb in a native or reduced state, and thus provide verification of amino acid composition and aid confirmation of protein sequence (Primary Structure chapter/Volume 2, Chapter 1). The technique can easily distinguish the different masses of the various glycoforms, thus providing information on the glycosylation profile, and the glycans can also be enzymatically removed so that the protein-only mass, as well as modifications such as glycation, can be determined. Light scattering techniques such as MALS (*68*) can also determine the MW of an antibody and any size variants present.

Charge

Deamidation and Isomerization

A key contributor to charge heterogeneity in mAbs is amino acid deamidation and isomerization (*5*). These are both forms of chemical modification and represent common degradation pathways for proteins *in vivo* (*69*) and in recombinant mAbs (*70–73*). Deamidation is a non-enzymatic process that typically occurs at Asn residues. Asn is converted to a 5-ringed cyclic succinimide intermediate that is hydrolyzed to form a mixture of iso-Asp and Asp in an approximate 3:1 ratio. Isomerization follows the same mechanism but occurs at Asp residues and proceeds through the succinimide intermediate to iso-Asp (*74*). Deamidation of glutamine (Gln) in a recombinant mAb has been reported, although it is kinetically less favorable than Asn deamidation due to the relative instability of its intermediates (*75*).

Several factors influence deamidation and isomerization rates, including primary structure, structural conformation, and the extrinsic environment. Studies on peptides have shown that the residue following Asn/Asp is an important determinant of the rate of deamidation and isomerization, with the sequences Asn-Gly and Asp-Gly being particularly susceptible to modification (*76, 77*). This is potentially due to the small Gly side chain not sterically constraining succinimide formation (*78*); however, other mechanisms have been suggested (*79*). If amino acid sequence "encodes" the modification potential, the realization of that potential is determined by structural conformation and the extrinsic environment. Secondary and tertiary structure can influence rates by determining the extent of solvent exposure and by facilitating interaction of Asn/Asp with proximal amino acids that can promote or prevent modification (*69, 76, 80*). Extrinsic factors such as pH, temperature and buffer species have all been shown to influence deamidation and isomerization (*78, 81*).

Deamidation and isomerization can have a significant impact on the development of a recombinant therapeutic by potentially influencing *in vitro* potency, product heterogeneity, shelf-life stability, manufacturing consistency, and yield. Of these, the principal concern is loss of therapeutic activity. Loss of *in vitro* activity in mAbs due to deamidation (*71, 72*) and isomerization (*74, 82*) has been reported. In all of these cases, the modification has occurred in the CDR, where it is presumed that changes in charge (deamidation) or structural configuration have perturbed antigen binding. Liu et al. showed *in vitro* deamidation rates of an Fc Asn were similar to *in vivo* rates (*69*); however, translation of CDR deamidation or isomerization to *in vivo* efficacy is still to be explored.

Due to the charge differences introduced by deamidation, it is readily monitored by charge-based methods such as cation exchange chromatography (CEX) or IEF (*74, 84*) (PTMs chapter/Volume 2, Chapter 3 and Separation chapter/Volume 2, Chapter 5). Isomerization has been resolved by CEX but is typically monitored by RP-HPLC and hydrophobic interaction chromatography (HIC) (*5, 80, 82, 83*). These methods exploit the differing hydrophobicity of isomerization products due to structural alterations caused by spatial changes in the side chain and the introduction of a methyl group into the peptide backbone (*74*). A drawback of all these methods is that they do not provide site-specific information. For this information, higher resolution methods such as peptide mapping with mass spectrometry are employed. Deamidation introduces a mass charge of +1 Dalton, which is readily detected by mass spectrometry (*72*) and can be localized to the specific residue modified. Distinguishing Asp from iso-Asp is typically achieved through chromatographic retention time because the two species have equal mass, although sophisticated mass spectrometry fragmentation techniques are a potential solution to this challenge (*85*). Iso-aspartyl bonds halt Edman degradation, enabling identification of iso-Asp sites, provided that the missing sequence information on the C-terminal side of the iso-Asp site can be obtained by other means.

The obvious approach to eliminating deamidation and isomerization is to engineer the Asn or Asp site to another amino acid residue. If the Asn/Asp is critical for maintaining activity, the adjacent amino acid can be mutated. For example, mutating the Asn carboxyl residue from Gly to a bulky hydrophobic residue dramatically reduces the propensity to deamidate (*77*). Other than molecular engineering, controlling the external environment of the molecule during manufacture, formulation, and storage are all vital for controlling Asn/Asp modification. For labile molecules, prolonged exposure to high pH during the manufacturing process should be avoided. Similarly, formulating at neutral pH with buffers and excipients that control solution properties and conformational flexibility should be evaluated (*78*). Lyophilization of the final drug product is often implemented for unstable molecules; however, deamidation has been shown to occur even in the solid state (*86*).

Sialylation

The overall charge distribution of a recombinant mAb may be influenced by the antibody glycosylation, and particularly, the degree of sialylation (Glycosylation chapter/Volume 2, Chapter 4). Both N- and O-linked oligosaccharides may be decorated with terminal sialic acid residues. Sialic acid residues are 9-carbon sugar units that are structurally related to neuraminic acid and carry a negative charge at physiological pH (*87*). If sialylation occurs on recombinant mAbs produced in mammalian cell lines, either *N*-glycolylneuraminic acid (NGNA) or *N*-acetylneuraminic acid (Neu5Ac) predominate. These sugars are both negatively charged but differ in structure, based on the nature of the chemical substituent at the 5-carbon position of the sugar ring.

The extent of sialylation can be influenced by the location of the carbohydrate attachment site and its relative accessibility on the mAb. For example, constant domain N297 glycans generally have a negligible to low level of sialylation, whereas variable domain glycans have a higher degree of sialic acid addition (*88, 89*).

Where sialylation occurs, the cell culture system used may affect the type and linkage of sialic acid added onto the protein. Golgi-resident, linkage-specific sialyltransferases transfer sialic acid residues from cytidine monophosphate (CMP)-sugar donors onto the terminal sugars of N- and O-glycans (*90*). Different cell culture systems have variation in the available CMP-sialic acid donor pools for sialylation. Consequently, murine cells lines such as NS0 generally decorate glycosylation with NGNA, and CHO cell line and human cell lines (e.g., HEK) with Neu5Ac (*91*). Human cell lines predominantly add Neu5Ac due to a mutation in the CMP-Neu5Ac hydroxylase gene that is required to produce CMP-NGNA (*92*). The exact nature of sialylation is not completely cell line-dependent; for example, CMP-NGNA may be generated in human cell lines by recycling NGNA from cell media, leading to NGNA sialylation on mAbs expressed in human cell lines (*93*). The carbohydrate linkage between the sialic acid and the glycan is based on the sialyltansferase complement of the cell as well. This is also distinct between cells lines, with α(2,6)-linked sialylation in human cells and α(2,3)-linked sialylation in CHO (*94*). Based on the sensitivity of sialylation to cell culture conditions, there is a significant body of work on the control of sialylation by modification of cell culture changes in both therapeutic mAbs (*95*) and non-immunoglobulin proteins (*96*).

Sialic acid may be measured indirectly as part of the global charge profile of the mAb, or directly as an individual feature by targeted methods. The impact that anionic sialic acid residues have on the overall charge pattern of a protein may be resolved as a series of bands or peaks on IEX and IEF methods, with each subsequent anionic species representing an additional sialic acid. The contribution of sialic acid to the global charge profile can be determined by comparing the charge profile of the mAb with and without enzymatic removal of sialic acid (*97*).

The quantity of sialic acid on a mAb may also be measured directly, which is generally achieved by complete removal of the sialic acid from the protein by acid or enzymatic hydrolysis, followed by quantification and characterization of the released sialic acid residues (*98*). The total release of sialic acid may be quantified by specific chemical reactions leading to a colorimetric or fluorescent end products (*99*). A commonly used approach to both identify and quantify released sialic acids is labeling the sugar with a fluorophore such as DMB (1,2-diamino-4,5-methylenoxybenzene) (*100*) or OPD (*o*-phenylenediamine) (*101*), and separating different species by RP chromatography, with quantification based on an external standard curve. Label-free quantification and characterization can be accomplished using high pH anion-exchange chromatography coupled to pulsed amperometric detection (*102*).

Where it occurs, sialylation can potentially affect the efficacy, immunogenicity, and pharmacokinetics of the mAb. Sialylation on fragment variable (Fv) glycans may be important for binding to the target antigen, affecting the efficacy of the protein (*103*). In non-immunoglobulins, sialylation is often a key determinant of the PK profile, acting as a cap on terminal galactose residues that are responsible for protein clearance via binding to the asialoglycoprotein receptor (*104*). Further, additional sialylation has also been shown to extend the PK half-life of smaller proteins (*105*). In mAbs, the impact of sialylation on pharmacokinetics is less clear. For example, both Millward et al. (*106*) and Huang et al. (*89*) demonstrate limited impact of variable domain sialylation on clearance in mouse models. There are, however, data for an impact of sialylated glycans on the route of clearance (*107*). Finally, the nature of sialylation may affect the immunogenic properties of the protein. As human proteins carry predominantly Neu5Ac, the presence of NGNA on therapeutic IgGs carries an increased risk of immunogenicity. Anti-NGNA antibodies have been observed in humans (*108, 109*), and these have been shown to bind to commercially available recombinant human mAbs (*91*). Taken together, this demonstrates the importance of monitoring and controlling the sialic acid quantity and composition of recombinant therapeutic antibody products.

Primary Structure and Post-Translational Modifications

Amino Acid Sequence Fidelity

As analytical methods continue to improve, we begin to find new sources of microheterogeneity within the light and heavy chains. Some of the variants are due to mis-incorporation, which originates when either the transfer RNA (tRNA) has been acylated with the incorrect amino acid or errors in translation have been made (Sequence Variant chapter/Volume 2, Chapter 2); both errors usually result in the appearance of the same replacement at multiple sites, such as Asn at Ser (*110*) or Ser at Asn (*111*). Mis-incorporation is detected using peptide mapping with mass spectrometry (*110, 111*).

Other sequence variants can result from mutations in the coding regions of the transfected heavy and light chain genes. These sequence variants generally appear as a single variation and can vary between clones (*112, 113*). The structural impacts of such sequence variants need to be assessed before a clone can be accepted for clinical or commercial production. For example, introduction or deletion of Pro or Cys residues, or the creation or deletion of an Asn-Xaa-Ser/Thr/Cys N-glycosylation site, may be considered a higher risk compared to conservative amino acid replacements. Sequence variants caused by mutations are usually detected by peptide mapping methods, and at high enough levels, these changes can also be seen by intact mass analysis, or even by SDS-PAGE (*114*). As our detection capabilities continue to improve, we will undoubtedly find more amino acid replacements, but the risk of low-level variants also has to be considered against a background of naturally occurring variants found at trace levels in non-recombinant (natural) proteins.

Disulfide Bonds

The disulfide bonding pattern for immunoglobulins is a well-established set of stable structures ((*115*); Figure 2 *and* 3). However, the presence of thioether (lanthionine; single sulfur) linkages and trisulfide bonds at what were expected to be disulfide linkages have been reported for recombinant antibodies (*116, 117*), generally at the interchain disulfide bond, sites such as the light chain Cys214 to heavy chain Cys220 bonds. The trisulfide forms have been attributed to overproduction of hydrogen sulfide in the bioreactor (*117*). Recent studies demonstrate thioether linkages form naturally over time *in vivo* for both endogenous and administered recombinant antibodies (*118*). IgG2 and IgG4 antibodies have distinct disulphide bond patterns to IgG1s. For IgG2s, heterogeneity at the interchain region has been observed, with this existing as a subset of isoforms designated IgG2-A, -B, and -A/B, (*119*). It has been demonstrated in a specific human IgG2 that these different disulfide isoforms have distinct activities in a cell-based assay. IgG4s can form monovalent half-molecules that exist in equilibrium with the bivalent *in vivo* (*120*). The expected disulfide bonding pattern of each subclass of IgG are listed in Figures 2 and 3.

Intrachain disulfide linkages may also be unoxidized (present as free thiols) (*121*), may affect bioactivity if the absence of the disulfide is in a constant domain 1 of an antibody heavy or light chain (C_H1 or C_L1) (*122*), and may be resolved chromatographically using HIC after papain cleavage into Fab and Fc fragments (*123*). Ellman's reagent may detect unpaired Cys residues only if the sample has been denatured, such as by incubation with chaotropic agents.

Glycation

Glycation is the result of reducing sugars, such as glucose, forming stable covalent modifications of Lys side chains or at the N-terminus. Glycation occurs with naturally occurring proteins, including antibodies (*124*). Glycated hemoglobin is a marker of diabetes, and degraded forms of glycated sites may also be recognized by receptors for advanced glycation end products. Recombinant proteins are produced in cell cultures that usually contain glucose, leading to multiple modifications of the same type, primarily of the Amadori product, which has an added 162 Dalton mass (*125*). Secondary structure can lead to unique sites being more extensively glycated (*126*). The Lys modification creates a reduced basic character; therefore, glycated sites are often enriched in acidic regions of charge profiles (*125*), and can often be detected using IEX and IEF methods in addition to mass spectrometry.

Glycosylation

Immunoglobulins contain a single N-glycosylation site at N297 in the heavy chain constant region and may contain additional N- or O-linked glycosylation sites in the variable domain on either the light or heavy chains. The glycosylation at N297 is predominantly bi-antennary, with different additions that include fucosylation, galactosylation, and low levels of sialylation. Additional glycosylation sites may occur in the variable domain of IgGs and will contain a distinct glycosylation pattern from N297, often with higher sialylation level. The presence of immature glycosylation variants, including high-mannose types, has also been observed at N-glycosylation sites. As a post-translational event occurring in the endo-membrane system of cells, the cell culture system and conditions may vary the structure and occupancy of glycosylation.

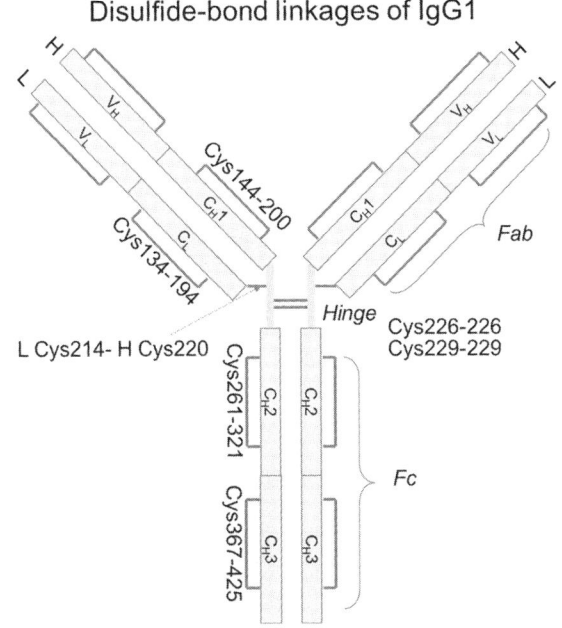

Figure 2. IgG1 and IgG4 disulfide bonding isoforms.

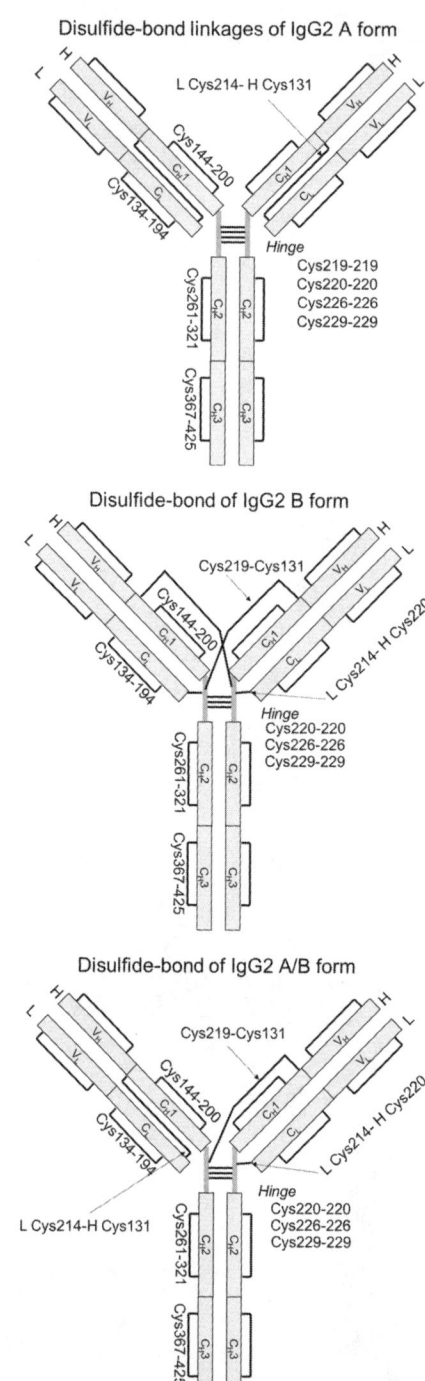

Figure 3. IgG2 disulfide bonding isoforms.

Analytical strategies for glycosylation determination may be tailored to the level of detail required (*127*), and are covered in-depth in the Glycosylation chapter (Volume 2, Chapter 4). The predominant glycosylation variants that are covalently attached to an intact IgG1 may be identified by intact protein mass spectrometry without deglycosylation; this approach is also useful for identifying additional non-Fc glycans (Intact chapter/Volume 3, Chapter 10). Site-specific analysis of glycopeptides by peptide mapping with LC-MS enables localization of glycosylation variants to a specific amino acid residue (*128*), and may be undertaken semiquantitatively to better understand the glycosylation distribution profile (*129*).

For detailed characterization, the most common approaches remove the glycan from its protein context by enzymatic release (notably, peptide N-glycosidase F [PNGaseF] and endoglycosidase H [Endo H]), and then characterize the oligosaccharides by a range of analytical techniques. For quantification, released oligosaccharides are typically labeled at the reducing terminus with a fluorophore such as anthranilic acid (2-AA) and 2-aminobenzamide (2-AB), and separated by chromatographic (*130*) or electrophoretic approaches (*131*). Peaks may then be structurally identified with reference to pure oligosaccharide standards by sequential enzymatic digestion with specific exo-glycanases or by on-line mass spectrometry. The identification of oligosaccharide without reducing-end labeling and separation is commonly undertaken by MALDI-TOF-MS using oligosaccharide-specific matrices (*132*).

Glycosylation has been demonstrated to have a significant impact on the safety and efficacy of antibody therapeutics. In terms of safety, specific glycosylation variants such as NGNA (*108, 109*) and terminal α-Gal may be immunogenic in humans (*133*). Glycosylation variants may affect the PK properties of the mAb, with high mannose variants being more rapidly cleared by receptor-mediated mechanisms (*134*). In terms of efficacy, glycosylation at N297 affects the local structure of the constant domain of the IgG and is critical for antibody effector function (*135*). Removal of glycosylation at N297 significantly diminishes FcγR- and C1q-mediated effector function mechanisms. Further, the structure of the glycan modulates the downstream effects of the mAb; for example, afucosylated glycans significantly increase antibody-dependent cell cytotoxicity (discussed in detail in the Mechanisms of Action chapter/Volume 1, Chapter 2). Based on the importance of glycosylation on antibody function, a number of strategies have been developed in cell culture to manipulate glycan structure, for example, engineering cell lines to produce afucosylated glycans (*136, 137*). In addition to the importance of Fc glycosylation, it has been demonstrated recently that engineering an oligosaccharide into the variable domain of an IgG may affect the activity of the antibody. For example, addition of a specific glycan onto ibalizumab increases *in vitro* neutralization of HIV-1 in resistant strains (*138*). Together, these examples show that both occupancy and structure of glycosylation within the constant and variable domains of an IgG may affect its clinical efficacy and safety.

Methionine and Tryptophan Oxidation

The oxidation of methionine (Met) or tryptophan (Trp) residues at critical sites can lead to a loss of potency or receptor binding. An early therapeutic antibody manufacturer used nitrogen in the headspace of the ampoule to reduce oxidative loss of potency for a mAb with a highly susceptible Met in a CDR (*139*). Met oxidation results in the addition of 16 Da and can affect RP-HPLC peptide map elution as well as hydrophobic interaction separations (*140*).

Similarly, Trp residues are susceptible to oxidation, leading to the formation of several products (*141*). Oxidation of a critical Trp residue led to the loss of potency for both a Fab product, ranibizumab (*142*), and a mAb (*143*). The effect of Fc Met oxidation on FcRn binding remains under investigation; the current literature indicates that an Fc needs to be oxidized at Met positions on both heavy chains to cause a binding decrease (*7, 144*). Absent any deleterious effects on potency or binding, demonstrating consistent low levels of oxidized forms can be useful to indicate the lack of impurities or atypical light exposure.

Conclusion

Three decades of recombinant protein product development have generated tools and techniques that enable the detailed characterization of therapeutic mAbs. Additional studies have determined the process origins and patient impacts of a myriad of structural variants. The regulatory credibility of a product sponsor depends on their ability to detect and control (by process controls or by QC testing) undesired variants, and to maintain a generally consistent analytical profile to ensure that undetected variants are also likely to be under control.

Earlier published work described common sources of single-site variation, such as heavy chain C-terminal Lys processing and deamidation. Size-based techniques enabled identification of fragmentation due to polypeptide chain cleavage, as well as the presence of dimeric or other aggregated forms. As our analytical capabilities have advanced, biopharmaceutical and academic laboratory groups continue to find new sources of variation, including Lys glycation, sequence variants, and variation at disulfide bond sites. Further analytical advances, along with improved biological characterization tools, will continue to increase our confidence that sponsors are producing safe and effective antibody-based therapeutics.

Acknowledgments

The authors are very grateful to Nancy Craighead for assistance on the preparation and review of this chapter. The authors gratefully acknowledge Jihong Wang for the disulfide bond drawing in Figures 2 and 3.

References

1. Chakhtoura, M.; Abdelnoor, A. M. *Immunopharmacol. Immunotoxicol.* **2010**, *32*, 533–542.
2. Coleman, K. *Datamonitor* **2011**.
3. Coleman, K. *Datamonitor* **2011**.
4. A-Mab: A Case Study in Bioprocess Development. http://c.ymcdn.com/sites/www.casss.org/resource/resmgr/imported/A-Mab_Case_Study_Version_2-1.pdf (accessed August 2014).
5. Harris, R. J.; Kabakoff, B.; Macchi, F. D.; Shen, F. J.; Kwong, M.; Andya, J. D.; Shire, S. J.; Bjork, N.; Totpal, K.; Chen, A. B. *J. Chromatogr. B: Biomed. Sci. Appl.* **2001**, *752*, 233–245.
6. Chirino, A. J.; Mire-Sluis, A. *Nat. Biotechnol.* **2004**, *22*, 1383–1391.
7. Wang, W.; Vlasak, J.; Li, Y.; Pristatsky, P.; Fang, Y.; Pittman, T.; Roman, J.; Wang, Y.; Prueksaritanont, T.; Ionescu, R. *Mol. Immunol.* **2011**, *48*, 860–866.
8. Beck, A.; Wagner-Rousset, E.; Ayoub, D.; Van Dorsselaer, A.; Sanglier-Cianferani, S. *Anal. Chem.* **2013**, *85*, 715–736.
9. Persson, K. M.; Gekas, V. *Process Biochem.* **1994**, *29*, 89–98.
10. Roberts, C. J.; Das, T. K.; Sahin, E. *Int. J. Pharm.* **2011**, *418*, 318–333.
11. Jing, Y.; Borys, M.; Nayak, S.; Egan, S.; Qian, Y.; Pan, S.; Li, Z. J. *Process Biochem.* **2012**, *47*, 69–75.
12. Elvin, J. G.; Couston, R. G.; van der Walle, C. F. *Int. J. Pharm.* **2013**, *440*, 83–98.
13. Stefani, M.; Dobson, C. M. *J. Mol. Med. (Berlin)* **2003**, *81*, 678–699.
14. Saluja, A.; Kalonia, D. S. *Int. J. Pharm.* **2008**, *358*, 1–15.
15. Philo, J. S.; Arakawa, T. *Curr. Pharm. Biotechnol.* **2009**, *10*, 348–351.
16. Mahler, H. C.; Friess, W.; Grauschopf, U.; Kiese, S. *J. Pharm. Sci.* **2009**, *98*, 2909–2934.
17. Narhi, L. O.; Schmit, J.; Bechtold-Peters, K.; Sharma, D. *J. Pharm. Sci.* **2012**, *101*, 493–498.
18. Cohen, S. I.; Vendruscolo, M.; Dobson, C. M.; Knowles, T. P. *J. Mol. Biol.* **2012**, *421*, 160–171.
19. Amani, S.; Naeem, A. *Process Biochem.* **2013**, *48*, 1651–1664.
20. Lumry, R.; Eyring, H. *J. Phys. Chem.* **1954**, *58*, 110–120.
21. Minton, A. P. *Biophys. J.* **2005**, *88*, 971–985.
22. Minton, A. P. *Curr. Opin. Struct. Biol.* **2000**, *10*, 34–39.
23. Breydo, L.; Reddy, K. D.; Piai, A.; Felli, I. C.; Pierattelli, R.; Uversky, V. N. *Biochim. Biophys. Acta* **2014**, *1844*, 346–357.
24. Bajaj, H.; Sharma, V. K.; Badkar, A.; Zeng, D.; Nema, S.; Kalonia, D. S. *Pharm. Res.* **2006**, *23*, 1382–1394.
25. Philo, J. S. *Curr. Pharm. Biotechnol.* **2009**, *10*, 359–372.
26. Lemmerer, M.; London, A. S.; Panicucci, A.; Gutierrez-Vargas, C.; Lihon, M.; Dreier, P. *J. Immunol. Methods* **2013**, *393*, 81–85.
27. Tatford, O. C.; Gomme, P. T.; Bertolini, J. *Biotechnol. Appl. Biochem.* **2004**, *40*, 67–81.
28. Philo, J. S. *AAPS J.* **2006**, *8*, E564–71.

29. Cao, S.; Pollastrini, J.; Jiang, Y. *Curr. Pharm. Biotechnol.* **2009**, *10*, 382–390.
30. Hamrang, Z.; Rattray, N. J.; Pluen, A. *Trends Biotechnol.* **2013**, *31*, 448–458.
31. Invernizzi, G.; Papaleo, E.; Sabate, R.; Ventura, S. *Int. J. Biochem. Cell Biol.* **2012**, *44*, 1541–1554.
32. Saguer, E.; Alvarez, P. A.; Sedman, J.; Ismail, A. A. *Food Hydrocolloids* **2013**, *33*, 402–414.
33. Kukrer, B.; Filipe, V.; van Duijn, E.; Kasper, P. T.; Vreeken, R. J.; Heck, A. J.; Jiskoot, W. *Pharm. Res.* **2010**, *27*, 2197–2204.
34. Wang, T.; Fodor, S.; Hapuarachchi, S.; Jiang, X. G.; Chen, K.; Apostol, I.; Huang, G. *J. Pharm. Biomed. Anal.* **2013**, *72*, 59–64.
35. Reschiglian, P.; Zattoni, A.; Roda, B.; Cinque, L.; Parisi, D.; Roda, A.; Dal Piaz, F.; Moon, M. H.; Min, B. R. *Anal. Chem.* **2005**, *77*, 47–56.
36. Chi, E. Y.; Krishnan, S.; Randolph, T. W.; Carpenter, J. F. *Pharm. Res.* **2003**, *20*, 1325–1336.
37. Rosenberg, A. S. *AAPS J.* **2006**, *8*, E501–7.
38. Braun, A.; Kwee, L.; Labow, M. A.; Alsenz, J. *Pharm. Res.* **1997**, *14*, 1472–1478.
39. Fradkin, A. H.; Carpenter, J. F.; Randolph, T. W. *J. Pharm. Sci.* **2009**, *98*, 3247–3264.
40. Hermeling, S.; Crommelin, D. J.; Schellekens, H.; Jiskoot, W. *Pharm. Res.* **2004**, *21*, 897–903.
41. Wang, X.; Das, T. K.; Singh, S. K.; Kumar, S. *mAbs* **2009**, *1*, 254–267.
42. Chennamsetty, N.; Helk, B.; Voynov, V.; Kayser, V.; Trout, B. L. *J. Mol. Biol.* **2009**, *391*, 404–413.
43. Shukla, D.; Schneider, C. P.; Trout, B. L. *Adv. Drug Delivery Rev.* **2011**, *63*, 1074–1085.
44. Abbas, S. A.; Sharma, V. K.; Patapoff, T. W.; Kalonia, D. S. *Pharm. Res.* **2012**, *29*, 683–694.
45. Harris, R. J. *J. Chromatogr. A* **1995**, *705*, 129–134.
46. Luo, J.; Zhang, J.; Ren, D.; Tsai, W. L.; Li, F.; Amanullah, A.; Hudson, T. *Biotechnol. Bioeng.* **2012**, *109*, 2306–2315.
47. Johnson, K. A.; Paisley-Flango, K.; Tangarone, B. S.; Porter, T. J.; Rouse, J. C. *Anal. Biochem.* **2007**, *360*, 75–83.
48. Tsubaki, M.; Terashima, I.; Kamata, K.; Koga, A. *Int. J. Biol. Macromol.* **2013**, *52*, 139–147.
49. Recny, M. A.; Scoble, H. A.; Kim, Y. *J. Biol. Chem.* **1987**, *262*, 17156–17163.
50. Fenn, J. B.; Mann, M.; Meng, C. K.; Wong, S. F.; Whitehouse, C. M. *Science* **1989**, *246*, 64–71.
51. Tanaka, K.; Waki, H.; Ido, Y.; Akita, S.; Yoshida, Y.; Yoshida, T.; Matsuo, T. *Rapid Commun. Mass Spectrom.* **1988**, *2*, 151–153.
52. Antes, B.; Amon, S.; Rizzi, A.; Wiederkum, S.; Kainer, M.; Szolar, O.; Fido, M.; Kircheis, R.; Nechansky, A. *J. Chromatogr. B: Anal. Technol. Biomed. Life Sci.* **2007**, *852*, 250–256.
53. Dick, L. W., Jr; Qiu, D.; Mahon, D.; Adamo, M.; Cheng, K. C. *Biotechnol. Bioeng.* **2008**, *100*, 1132–1143.

54. Khawli, L. A.; Goswami, S.; Hutchinson, R.; Kwong, Z. W.; Yang, J.; Wang, X.; Yao, Z.; Sreedhara, A.; Cano, T.; Tesar, D.; Nijem, I.; Allison, D. E.; Wong, P. Y.; Kao, Y. H.; Quan, C.; Joshi, A.; Harris, R. J.; Motchnik, P. *mAbs* **2010**, *2*, 613–624.
55. Meert, C. D.; Brady, L. J.; Guo, A.; Balland, A. *Anal. Chem.* **2010**, *82*, 3510–3518.
56. Alexander, A. J.; Hughes, D. E. *Anal. Chem.* **1995**, *67*, 3626–3632.
57. Cordoba, A. J.; Shyong, B. J.; Breen, D.; Harris, R. J. *J. Chromatogr. B: Anal. Technol. Biomed. Life Sci.* **2005**, *818*, 115–121.
58. Xiang, T.; Lundell, E.; Sun, Z.; Liu, H. *J. Chromatogr. B: Anal. Technol. Biomed. Life Sci.* **2007**, *858*, 254–262.
59. Dillon, T. M.; Bondarenko, P. V.; Rehder, D. S.; Pipes, G. D.; Kleemann, G. R.; Ricci, M. S. *J. Chromatogr. A* **2006**, *1120*, 112–120.
60. Cohen, S. L.; Price, C.; Vlasak, J. *J. Am. Chem. Soc.* **2007**, *129*, 6976–6977.
61. Smith, M. A.; Easton, M.; Everett, P.; Lewis, G.; Payne, M.; Riveros-Moreno, V.; Allen, G. *Int. J. Pept. Protein Res.* **1996**, *48*, 48–55.
62. Ouellette, D.; Alessandri, L.; Piparia, R.; Aikhoje, A.; Chin, A.; Radziejewski, C.; Correia, I. *Anal. Biochem.* **2009**, *389*, 107–117.
63. Vlasak, J.; Ionescu, R. *mAbs* **2011**, *3*, 253–263.
64. Liu, H.; Gaza-Bulseco, G.; Lundell, E. *J. Chromatogr. B: Anal. Technol. Biomed. Life Sci.* **2008**, *876*, 13–23.
65. Ishikawa, T.; Ito, T.; Endo, R.; Nakagawa, K.; Sawa, E.; Wakamatsu, K. *Biol. Pharm. Bull.* **2010**, *33*, 1413–1417.
66. Sandberg, H.; Lutkemeyer, D.; Kuprin, S.; Wrangel, M.; Almstedt, A.; Persson, P.; Ek, V.; Mikaelsson, M. *Biotechnol. Bioeng.* **2006**, *95*, 961–971.
67. Gao, S. X.; Zhang, Y.; Stansberry-Perkins, K.; Buko, A.; Bai, S.; Nguyen, V.; Brader, M. L. *Biotechnol. Bioeng.* **2011**, *108*, 977–982.
68. Kendrick, B. S.; Kerwin, B. A.; Chang, B. S.; Philo, J. S. *Anal. Biochem.* **2001**, *299*, 136–146.
69. Liu, Y. D.; van Enk, J. Z.; Flynn, G. C. *Biologicals* **2009**, *37*, 313–322.
70. Chelius, D.; Rehder, D. S.; Bondarenko, P. V. *Anal. Chem.* **2005**, *77*, 6004–6011.
71. Vlasak, J.; Bussat, M. C.; Wang, S.; Wagner-Rousset, E.; Schaefer, M.; Klinguer-Hamour, C.; Kirchmeier, M.; Corvaia, N.; Ionescu, R.; Beck, A. *Anal. Biochem.* **2009**, *392*, 145–154.
72. Huang, L.; Lu, J.; Wroblewski, V. J.; Beals, J. M.; Riggin, R. M. *Anal. Chem.* **2005**, *77*, 1432–1439.
73. Schenerman, A.; Axley, M.; Oliver, C.; Ram, K.; Wasserman, G. In Using a Risk Assessment Process to Determine Criticality of Product Quality Attributes. In *Quality by Design for Biopharmaceuticals: Principles and Case Studies*; Rathore, A. S., Mahtre R. Eds.; John Wiley & Sons, Inc.: Hoboken, NJ, 2009.
74. Vlasak, J.; Ionescu, R. *Curr. Pharm. Biotechnol.* **2008**, *9*, 468–481.
75. Liu, H.; Gaza-Bulseco, G.; Chumsae, C. *Rapid Commun. Mass Spectrom.* **2008**, *22*, 4081–4088.
76. Wakankar, A. A.; Borchardt, R. T.; Eigenbrot, C.; Shia, S.; Wang, Y. J.; Shire, S. J.; Liu, J. L. *Biochemistry* **2007**, *46*, 1534–1544.

77. Robinson, N. E. *Proc. Natl. Acad. Sci. U.S.A.* **2002**, *99*, 5283–5288.
78. Wakankar, A. A.; Borchardt, R. T. *J. Pharm. Sci.* **2006**, *95*, 2321–2336.
79. Radkiewicz, J. L.; Zipse, H.; Clarke, S.; Houk, K. N. *J. Am. Chem. Soc.* **2001**, *123*, 3499–3506.
80. Cacia, J.; Keck, R.; Presta, L. G.; Frenz, J. *Biochemistry* **1996**, *35*, 1897–1903.
81. Pace, A. L.; Wong, R. L.; Zhang, Y. T.; Kao, Y. H.; Wang, Y. J. *J. Pharm. Sci.* **2013**, *102*, 1712–1723.
82. Rehder, D. S.; Chelius, D.; McAuley, A.; Dillon, T. M.; Xiao, G.; Crouse-Zeineddini, J.; Vardanyan, L.; Perico, N.; Mukku, V.; Brems, D. N.; Matsumura, M.; Bondarenko, P. V. *Biochemistry* **2008**, *47*, 2518–2530.
83. Xiao, G.; Bondarenko, P. V. *J. Pharm. Biomed. Anal.* **2008**, *47*, 23–30.
84. Perkins, M.; Theiler, R.; Lunte, S.; Jeschke, M. *Pharm. Res.* **2000**, *17*, 1110–1117.
85. Yang, H.; Zubarev, R. A. *Electrophoresis* **2010**, *31*, 1764–1772.
86. Lai, M. C.; Topp, E. M. *J. Pharm. Sci.* **1999**, *88*, 489–500.
87. Varki, A.; Schauer, R. In *Sialic Acids*; Varki, A., Cummings, R. D., Esko, J. D., Freeze, H. H., Stanley, P., Bertozzi, C. R., Hart, G. W.; Etzler, M. E., Eds.; *Essentials of Glycobiology*; Cold Spring Harbor: New York, 2009
88. Raju, T. S.; Jordan, R. E. *mAbs* **2012**, *4*, 385–391.
89. Huang, L.; Biolsi, S.; Bales, K. R.; Kuchibhotla, U. *Anal. Biochem.* **2006**, *349*, 197–207.
90. Harduin-Lepers, A.; Vallejo-Ruiz, V.; Krzewinski-Recchi, M. A.; Samyn-Petit, B.; Julien, S.; Delannoy, P. *Biochimie* **2001**, *83*, 727–737.
91. Ghaderi, D.; Taylor, R. E.; Padler-Karavani, V.; Diaz, S.; Varki, A. *Nat. Biotechnol.* **2010**, *28*, 863–867.
92. Varki, A. *Nature* **2007**, *446*, 1023–1029.
93. Bardor, M.; Nguyen, D. H.; Diaz, S.; Varki, A. *J. Biol. Chem.* **2005**, *280*, 4228–4237.
94. Bragonzi, A.; Distefano, G.; Buckberry, L. D.; Acerbis, G.; Foglieni, C.; Lamotte, D.; Campi, G.; Marc, A.; Soria, M. R.; Jenkins, N.; Monaco, L. *Biochim. Biophys. Acta* **2000**, *1474*, 273–282.
95. Hossler, P.; Khattak, S. F.; Li, Z. J. *Glycobiology* **2009**, *19*, 936–949.
96. Jing, Y.; Qian, Y.; Li, Z. J. *Biotechnol. Bioeng.* **2010**, *107*, 488–496.
97. Sundaram, S.; Matathia, A.; Qian, J.; Zhang, J.; Hsieh, M. C.; Liu, T.; Crowley, R.; Parekh, B.; Zhou, Q. *mAbs* **2011**, *3*, 505–512.
98. Reuter, G.; Schauer, R. *Methods Enzymol.* **1994**, *230*, 168–199.
99. Markely, L. R.; Ong, B. T.; Hoi, K. M.; Teo, G.; Lu, M. Y.; Wang, D. I. *Anal. Biochem.* **2010**, *407*, 128–133.
100. Hara, S.; Takemori, Y.; Yamaguchi, M.; Nakamura, M.; Ohkura, Y. *Anal. Biochem.* **1987**, *164*, 138–145.
101. Anumula, K. R. *Anal. Biochem.* **1995**, *230*, 24–30.
102. Rohrer, J. S.; Thayer, J.; Weitzhandler, M.; Avdalovic, N. *Glycobiology* **1998**, *8*, 35–43.
103. Khurana, S.; Raghunathan, V.; Salunke, D. M. *Biochem. Biophys. Res. Commun.* **1997**, *234*, 465–469.

104. Webster, R.; Taberner, J.; Edgington, A.; Guhan, S.; Varghese, J.; Feeney, H.; Blocker, L.; Jezequel, S. G. *Xenobiotica* **1999**, *29*, 1141–1155.
105. Egrie, J. C.; Dwyer, E.; Browne, J. K.; Hitz, A.; Lykos, M. A. *Exp. Hematol.* **2003**, *31*, 290–299.
106. Millward, T. A.; Heitzmann, M.; Bill, K.; Langle, U.; Schumacher, P.; Forrer, K. *Biologicals* **2008**, *36*, 41–47.
107. Wright, A.; Sato, Y.; Okada, T.; Chang, K.; Endo, T.; Morrison, S. *Glycobiology* **2000**, *10*, 1347–1355.
108. Tangvoranuntakul, P.; Gagneux, P.; Diaz, S.; Bardor, M.; Varki, N.; Varki, A.; Muchmore, E. *Proc. Natl. Acad. Sci. U.S.A.* **2003**, *100*, 12045–12050.
109. Padler-Karavani, V.; Yu, H.; Cao, H.; Chokhawala, H.; Karp, F.; Varki, N.; Chen, X.; Varki, A. *Glycobiology* **2008**, *18*, 818–830.
110. Yu, X. C.; Borisov, O. V.; Alvarez, M.; Michels, D. A.; Wang, Y. J.; Ling, V. *Anal. Chem.* **2009**, *81*, 9282–9290.
111. Wen, D.; Vecchi, M. M.; Gu, S.; Su, L.; Dolnikova, J.; Huang, Y. M.; Foley, S. F.; Garber, E.; Pederson, N.; Meier, W. *J. Biol. Chem.* **2009**, *284*, 32686–32694.
112. Harris, R. J.; Murnane, A. A.; Utter, S. L.; Wagner, K. L.; Cox, E. T.; Polastri, G. D.; Helder, J. C.; Sliwkowski, M. B. *Biotechnology (N.Y.)* **1993**, *11*, 1293–1297.
113. Guo, D.; Gao, A.; Michels, D. A.; Feeney, L.; Eng, M.; Chan, B.; Laird, M. W.; Zhang, B.; Yu, X. C.; Joly, J.; Snedecor, B.; Shen, A. *Biotechnol. Bioeng.* **2010**, *107*, 163–171.
114. Wan, M.; Shiau, F. Y.; Gordon, W.; Wang, G. Y. *Biotech Bioeng.* **1999**, *62*, 485–488.
115. Wypych, J.; Guo, A.; Zhang, Z.; Martinez, T.; Allen, M. J.; Fodor, S.; Kelner, D. N.; Flynn, G. C.; Liu, Y. D.; Bondarenko, P. V.; Ricci, M. S.; Dillon, T. M.; Balland, A. J. *J. Biol. Chem* **2008**, *283*, 16194–16205.
116. Tous, G. I.; Wei, Z.; Feng, J.; Bilbulian, S.; Bowen, S.; Smith, J.; Strouse, R.; McGeehan, P.; Casas-Finet, J.; Schenerman, M. A. *Anal. Chem.* **2005**, *77*, 2675–2682.
117. Gu, S.; Wen, D.; Weinreb, P. H.; Sun, Y.; Zhang, L.; Foley, S. F.; Kshirsagar, R.; Evans, D.; Mi, S.; Meier, W.; Pepinsky, RB *Anal Biochem.* **2010**, *400*, 89–98.
118. Zhang, Q.; Schenauer, M. R.; McCarter, JD.; Flynn, G. C. *J. Biol. Chem.* **2013**, *288*, 16371–16382.
119. Dillon, T. M.; Ricci, M. S.; Vezina, C.; Flynn, G. C.; Liu, Y. D.; Rehder, D. S.; Plant, M.; Henkle, B.; Li, Y.; Deechongkit, S.; Varnum, B.; Wypych, J.; Balland, A.; Bondarenko, P. V. *J. Biol. Chem.* **2008**, *283*, 16206–16215.
120. Salfeld, J. G. *Nat. Biotechnol.* **2007**, *25*, 1369–1372.
121. Zhang, W.; Czupryn, M. J. *Biotechnol. Prog.* **2002**, *18*, 509–513.
122. Harris, R. J.; Chin, E. T.; Macchi, F.; Keck, R. G.; Shyong, B.; Ling, V. T.; Cordoba, A. J.; Marian, M.; Sinclair, D.; Battersby, J. E.; Jones, A. J. S. Analytical Characterization of Monoclonal Antibodies: Linking Structure to Function; In *Current Trends in Monoclonal Antibody Development and Manufacturing*; Shire, S. J., Gombotz, W., Bechtold-Peters, K., Andya, J., Eds.; Springer: New York, 2010; pp 193–205.

123. Chaderjian, W. B.; Chin, E. T.; Harris, R. J.; Etcheverry, T. M. *Biotechnol. Prog.* **2005**, *21*, 550–553.
124. Lapolla, A.; Fedele, D.; Garbeglio, M.; Martano, L.; Tonani, R.; Seraglia, R.; Favretto, D.; Fedrigo, M. A.; Traldi, P. *J. Am. Soc. Mass Spectrom.* **2000**, *11*, 153–159.
125. Quan, C.; Alcala, E.; Petkovska, I.; Matthews, D.; Canova-Davis, E.; Taticek, R.; Ma, S. *Anal. Biochem.* **2008**, *373*, 179–191.
126. Zhang, B.; Yang, Y.; Yuk, I.; Pai, R.; McKay, P.; Eigenbrot, C.; Dennis, M.; Katta, V.; Francissen, K. C. *Anal. Chem.* **2008**, *80*, 2379–2390.
127. Hunh, C.; Selman, M. H. J.; Ruhaak, R.; Deelder, A. M.; Wuhrer, M. *Proteomics* **2009**, *9*, 882–913.
128. Zauner, G.; Selman, M. H.; Bondt, A.; Rombouts, Y.; Blank, D.; Deelder, A. M.; Wuhrer, M. *Mol. Cell. Proteomics* **2013**, *12*, 856–865.
129. Toyama, A.; Nakagawa, H.; Matsuda, K.; Sato, T. A.; Nakamura, Y.; Ueda, K. *Anal Chem.* **2012**, *20*, 9655–9662.
130. Melmer, M.; Strangler, T.; Schiefermeier, M.; Brunner, W.; Toll, T.; Rupprechter, A.; Linder, W.; Premstaller, A. *Anal. Bioanal. Chem.* **2010**, *398*, 905–914.
131. Kamoda, S.; Nomura, C.; Kinoshita, M.; Nishiura, S.; Ishikawa, R.; Kakehi, K.; Kawasaki, N.; Hayakawa, T. *J. Chromatogr. A* **2004**, *1050*, 211–216.
132. Qian, J.; Liu, T.; Yang, L.; Daus, A.; Crowley, R.; Zhou, Q. *Anal. Biochem.* **2007**, *364*, 8–18.
133. Macher, B. A.; Galili, U. *Biochim. Biophys. Acta* **2008**, *1780*, 75–88.
134. Yu, M; Brown, D; Reed, C; Chung, S; Lutman, J; Stefanich, E; Wong, A; Stephan, JP; Bayer, R *mAbs* **2012**, *4*, 475–487.
135. Jefferis, R. *Trends Pharmacol. Sci.* **2009**, *30*, 356–362.
136. Umana, P.; Mairet, J. J.; Moudry, R.; Amusutzl, H.; Bailey, J. E. *Nat. Biotechnol.* **2002**, *277*, 26733–26740.
137. Shields, R. L.; Lai, J.; Keck, R.; O'Connell, L. Y.; Hong, K.; Meng, M. Y.; Weikert, S. H.; Presta, L. G. *J. Biol. Chem.* **2002**, *277*, 26733–26740.
138. Song, R.; Oren, D. A.; Franco, D.; Seaman, M.; Ho, D. D. *Nat. Biotechnol.* **2013**, *31*, 1047–1052.
139. Kroon, D. J.; Baldwin-Ferro, A.; Lalan, P. *Pharm. Res.* **1992**, *9*, 1386–1393.
140. Shen, F. J.; Kwong, M. Y.; Keck, R. G.; Harris, R. J. In *Techniques in Protein Chemistry VII*; Marshak, D. R., Ed.; Academic Press: Waltham, MA, 1996; pp 275–284.
141. Taylor, S. W.; Fahy, E.; Murray, J.; Capaldi, R. A.; Ghosh, S. S. *J. Biol. Chem.* **2003**, *278*, 19587–19590.
142. Lam, X. M.; Lai, W. G.; Chan, E. K.; Ling, V.; Hsu, C. C. *Pharm. Res.* **2011**, *28*, 2543–2555.
143. Wei, Z.; Feng, J.; Lin, H. Y.; Mullapudi, S.; Bishop, E.; Tous, G. I.; Casas-Finet, J.; Hakki, F.; Strouse, R.; Schenerman, M. A. *Anal. Chem.* **2007**, *79*, 2797–2805.
144. Schlothauer, T.; Rueger, P.; Stracke, J. O.; Hertenberger, H.; Fingas, F.; Kling, L.; Emrich, T.; Drabner, G.; Seeber, S.; Auer, J.; Koch, S.; Papadimitriou, A. *mAbs* **2013**, *5*, 576–586.

Chapter 4

Perspectives on Well-Characterized Biological Proteins

Kurt Brorson[*,1] and Brent Kendrick[2]

[1]Center for Drug Evaluation and Research,
U.S. Food and Drug Administration,
Silver Spring, Maryland 20903, United States
[2]Amgen Inc., Longmont, Colorado 80503, United States
*E-mail: kurt.brorson@fda.hhs.gov

The public expects medicines, including biopharmaceuticals, to be pure, of high quality, and consistent between doses. A critical component of achieving this goal is establishment and control of the critical quality attributes (CQAs) of both the bulk protein drug substance and the final unit dosage form that is administered to the patient. A well-characterized biological protein is one where there is confidence that all features important for product safety and potency have been identified. In this chapter, an industry and a regulatory authority representative discuss perspectives on achieving the shared goal of making high-quality biopharmaceuticals available to the public.

Introduction

ICH Q8(R2) defines a critical quality attribute (CQA) as "a physical, chemical, biological, or microbiological property or characteristic that should be within an appropriate limit, range, or distribution to ensure the desired product quality." Product heterogeneity is an unavoidable aspect of biotechnology products; thus as specified by ICH Q6B, the degree and profile of the heterogeneity should be characterized and shown to be consistent. The imperative to gather product attribute information supports the understanding of the overall product biochemistry and the criticality of each attribute, leading to safe use and consistent benefit to patients.

As stated above, complex biotechnology products do not consist of a single biochemical entity but rather are heterogeneous, containing product-related variants and process-related impurities. An understanding of range and distribution of product quality attributes (PQAs), including their physicobiochemical characterization, as well as the impact of potential variants on safety and efficacy, should be a core component of product development during the entire product lifecycle.

Defining CQAs of a biotechnology product is a complex task, and there probably never will be a single list of CQAs or a single method for CQA identification or ranking. Quality by design (QbD) approaches have helped in defining this process, but as noted in ICH Q9, even in the case of risk assessments, no single method is applicable in all cases. ICH Q5E states that in cases where differences are seen between products in a comparability study, the "… existing knowledge [needs to be] sufficiently predictive to ensure that any differences in quality attributes have no adverse impact upon safety and/or efficacy." Thus, a complete understanding of the science underpinning the quality attributes is critical and would inform a risk-based approach for decision making.

Beyond routine testing, experience by industry and regulators has revealed some more subtle variants and areas of potential focus. Although not meant to be an exhaustive list, the following may apply to individual products on a case-by-case basis:

- Reduction. The occurrence of reduction of the interchain disulfide bonds in monoclonal antibody (mAb) products at the time of bioreactor harvest or storage of the harvest material, primarily due to mechanical shear-induced lysis of viable cells, which results in the release of enzymes that mediate disulfide bond reduction. Although in many cases, noncovalent interactions may be strong enough to hold molecules intact, disulfide scrambling can lead to subsequent aggregation or subtle misfolding.
- IgG2 and IgG4 Complexity. IgG2 antibodies can exist as a mixture of disulfide-linked structural isoforms that can have different binding affinities and potencies. IgG4 antibodies have the ability to form half antibodies and exchange Fab arms with other IgG4 antibodies, resulting in bispecific, monovalent antibodies.
- Product misfolding. This aspect is particularly important for Fc fusion proteins and other molecules that do not have a "native" counterpart. Methods capable of evaluating both secondary and tertiary structure need to be employed in these cases.
- Chemical modifications of select side chains, including glycation. These chemical reactions can occur both during protein production in the cells and later, for example, in the presence of high sugar concentration formulation buffers.
- Product aggregates. Aggregates have the potential to affect both the safety and efficacy of a product through alteration of the potency and/or immunogenicity. Aggregates can form during almost any stage of manufacturing, as well as during storage and shipment. Protein aggregates can be reversible or irreversible, soluble or insoluble,

and homo- or heteronucleated. Aggregate sizes can be grouped into submicron aggregates (high molecular weight [HMW] species), nanometer aggregates (oligomers), micron-scale aggregates (subvisible particles), and aggregates greater than 100 micron (visible aggregate). For additional discussion, see the Aggregation chapter/Volume 3, Chapter 5 and Protein Particulates chapter/Volume 2, Chapter 8.

Analytical Techniques

The utility of an assay depends in large part on its range, accuracy, precision, specificity, and sensitivity. ICH Q2(R1) provides specific information concerning what parameters to evaluate during an assay validation. Recently, QbD-type approaches have been proposed for assay validation, for example, applying design of experiments (DoE) matrices to robustness studies. In addition, consideration should also be given as to whether the analytical methods that are used are appropriate for their intended use and whether any new analytical tools will be needed. Analytical programs have at least three levels: (1) release/in-process control testing, (2) characterization/comparability, and (3) stability. The range of analytical assays should be tailored to the specific needs of each level and each product.

- Release (and routine in-process testing). Depending on the extent of the analytical testing performed for each batch of drug substance and drug product, as well as for routine in-process samples, the assays provide information on product attributes but generally are not expected to comprehensively characterize the full biochemistry of the product or determine comparability.
- Extended physicochemical and biological characterization. Comparability (between process evolutions) and characterization (of the original protein molecule) studies include at a minimum the release analytical testing plus a more extended analysis that comprehensively characterizes the biochemistry and other aspects of the product. The number and type of extended characterization methods used will be dependent on the extent of the molecular complexity, the process, and the phase of development. Ideally, a comparability study will employ methods that use orthogonal physicochemical and biological principles to analyze quality attributes to maximize product understanding.
- Stability. Stability studies are an important aspect of product understanding. Changes over time should be minimized, and if changes do occur, data or a risk assessment demonstrating that the change has not adversely impacted product safety or function is warranted. Certain assays that detect product instability prior to others are referred to as "stability indicating assays." These should be identified for each product at some point in development.

New Analytical Methods

As stated in ICH Q6B, the composition, physical properties, potency, and primary structure of a biotechnology product should be characterized to the extent feasible. Because of the inherent heterogeneity of these products, a battery of techniques that employ different physiochemical or biological principles is warranted for this purpose. Over time, new analytical techniques become available that can provide enhancements in parameters such as sensitivity, specificity, linearity, and reproducibility. Some new assays also possess more robust performance and can tolerate variations such as buffer components or sample volumes, making these amenable for multiproduct use. In addition, advancements in post-separation detection methods have enabled an increase in the sensitivity of many analytical techniques. Many of the newer methods are subjects of subsequent chapters in this book and can be applied as appropriate.

When safety or efficacy concerns surrounding a particular product class are identified based on new information, developing or updating methods is warranted. An example of this is the recent attention to 2–10 μM sized aggregate formation, which can be evaluated with non-routine methods such as subvisible particle imaging techniques, multi-angle light scattering (MALS), and/or field flow fractionation (FFF). A need to update or change methods also can arise for some legacy or less commonly used technologies or as certain reagents lose vendor support or become scarce (e.g., acetonitrile, ampholytes, protein markers). Implementation of new or modified analytical technology is even encouraged by ICH Q6B (ICH quality guidelines) and generally should be encouraged by regulatory agencies. When the new method is intended to replace an existing method, information and data that demonstrate that the new method has the same or better performance capabilities compared to the original method should be made available to regulatory authorities.

Common Strategies for Structural and Biochemical Characterization

Biochemical, biophysical, and biological characterization of a protein therapeutic is conducted during clinical development to provide a comprehensive understanding of its structural and functional properties; enable assessment of the criticality of PQAs; and, ultimately, provide the foundation for product control strategy (specification, comparability and stability strategies) per ICH Q8, ICH Q6B, ICH Q9, ICH Q5E, ICH Q2, ICH Q1A(R2) and ICH Q5C (*1–7*). The extent of product characterization will vary depending on the clinical phase of the product, with the most comprehensive assessment (and the focus of this paper) occurring in later stages and submitted in the Marketing Application in the Elucidation of Structure section. All elucidation of structure studies for the Marketing Application are typically conducted using material representative of the commercial process at commercial scale. Drug substance and drug product may be considered interchangeable for the purpose of structural elucidation unless scientific considerations render them non-interchangeable.

A summary of the techniques typically used for these characterization studies is presented in Table 1. All methods are typically qualified as fit for purpose. An in-depth discussion and representative data for the NISTmAb using many of these characterization techniques can be found throughout this book series.

Table 1. Example Characterization Method Summary

Category	Purpose	Technique
Primary Structure	Mass confirmation based on sequence, assessment of mass variants	Native, deglycosylated, and reduced protein electrospray ionization-mass spectrometry (ESI-MS)
	Amino acid sequence confirmation	MS/MS sequencing of peptide maps
	N- and C-terminal variants	MS/MS sequencing of peptide maps
	Other post-translational modifications	MS/MS sequencing of peptide maps
Glycosylation	N-linked glycosylation	Peptide or glycan map with MS, permethylation with MSn
	O-linked glycosylation	Peptide map with MS
	Attribute criticality	Enzymatic digest, bioassay
Disulfide Structure	Identification of disulfide-linked peptides	Non-reduced and reduced peptide map with MS
	Free sulfhydryls	Ellman's assay (or similar)
Charge Variants	Charge heterogeneity	Cation exchange high-performance liquid chromatography (CEX-HPLC)
	Charge heterogeneity	Capillary isoelectric focusing (cIEF)
	Attribute criticality	Purification and characterization of charge variants
Size Variants	Size heterogeneity under native conditions	Size exclusion chromatography (SEC)
	Size heterogeneity of reducible forms	Reduced capillary sodium dodecylsulfate electrophoresis (rcSDS); reduced, denatured SEC (rdSEC)
	Size heterogeneity of covalent forms	nrcSDS, dSEC
	Determination of monomer, dimer, and submicron aggregates	Resolution by SEC with static light scattering detection (SEC-SLS)
	Size heterogeneity and sedimentation coefficient determination	Sedimentation velocity analytical ultracentrifugation (SV-AUC)

Continued on next page.

Table 1. (Continued). Example Characterization Method Summary

Category	Purpose	Technique
	Attribute criticality	Purification and characterization of size variants
Biophysical Characterization	Secondary structure	Fourier transform infrared (FTIR) spectroscopy
	Tertiary structure	Near-UV circular dichroism (near-UV CD) spectroscopy
	Thermal stability	Differential scanning calorimetry (DSC)
Biological Characterization	Justification of bioassay relevance to mechanism of action (MOA)	Bioassay(s)
	Probe relevant functional domains	Receptor binding assays
	Assess other relevant biological functions (e.g., complement-dependent cytotoxicity [CDC], antibody-dependent cellular cytotoxicity [ADCC])	Other characterization biological assays

Primary Structure

The amino acid sequence of the product is determined by the gene sequence and the fidelity of transcription of the complementary DNA (cDNA) gene cloned into the cell line. The gene sequence can be confirmed by generating reverse transcription polymerase chain reaction (RT PCR) of messenger RNA (mRNA) isolated from the cells and subsequently amplified using PCR to generate sufficient material for direct sequencing. The presence of an intact gene transcript of the correct molecular weight can be further confirmed by Northern blotting, and the integrity of the product genes can be analyzed by Southern blotting to detect any gross insertions, deletions, or rearrangements.

Upon confirmation that the correct gene sequence is cloned into the cell line, the primary structure (amino acid sequence) of the drug substance can be confirmed using a combination of whole (intact) molecule mass spectrometry (MS) and peptide mapping analysis (Primary Structure chapter/Volume 2, Chapter 1).

Whole mass analysis of native product and the constituent components (e.g., heavy chain [H] and light chain [L] in mAb's) serve to confirm the identity and integrity of the protein. This can be accomplished by comparing the theoretical calculated masses based on amino acid sequence from the cDNA gene and accounting for the presence of post-translational modifications (including glycosylation) to the experimentally measured masses of intact, deglycosylated, and reduced and deglycosylated forms. MS and MS/MS of proteolytic digests can be further used to establish amino acid sequence. The complete sequence of every individual amino acid may not be practical in some cases (e.g., large proteolytic fragments can evade full MS/MS sequencing) and should be determined to the extent practical (2).

Sequence variant analysis should also be performed (Sequence Variant chapter/Volume 2, Chapter 2). Proteins are often subject to several post-translational modifications that result in heterogeneity during expression in the cell culture process (PTMs chapter/Volume 2, Chapter 3). For example, N-terminal variants may occur due to cyclization of the first residue (8). Additionally, N-terminal acetylation may occur after cleavage of the N-terminal methionine by methionine aminopeptidase and replacing the amino acid with an acetyl group from acetyl-CoA by N-acetyltransferase enzymes (9).

C-terminal variants are known to arise during cell culture from removal of C-terminal Lys residues due to post-translational carboxypeptidase processing by enzymes such as peptidylglycine α-amidating monooxygenase (10). Additionally, amidation of the penultimate amino acid may occur with the sequence R-Xxx-Gly, where R is the main body of the protein, and Xxx is the residue that is amidated. The Gly residue is cleaved, donating the amino moiety to the penultimate amino acid, resulting in C-terminal amidation (11, 12).

Methionine oxidation is also a common modification occurring during production or storage. Oxidation of other residues is also possible, especially tryptophan and histidine (13). Oxidation stress can occur from a number of sources such as exposure to air, light, peroxy radicals (especially those coming from polysorbate formulation components), and/or trace metals (14). Forced oxidation using UV light, visible light, and hydrogen or tert-butyl peroxide can also help elucidate the susceptibility to oxidation at specific sites on the molecule, and bioassay analysis of oxidized samples can help establish any potential product quality impacts of oxidation.

Asparagine deamidation is another common modification. Similar to oxidation assessment, forced deamidation under high pH conditions can help establish susceptibility of specific sites and any impact on potency.

Glycosylation

Glycoproteins (both naturally occurring *in vivo* as well as recombinant) typically contain various glycoforms (15). It is difficult to cover all conceivable glycoforms here because glycosylation is a function of the host cell/expression system, number and locations of glycosylation sites, and numerous other factors. The diversity of glycans include asialylated and sialylated forms, various antennary structures, core-fucosylated and afucosylated structures, and other variants.

Glycosylation characterization studies are performed to identify the site(s) and percentage occupancy of *N*-linked glycosylation sites, characterize the associated *N*-glycans, and identify the presence of any *O*-linked glycans (Glycosylation chapter/Volume 2, Chapter 4). Glycosylation may be characterized using a variety of techniques, including (16, 17):

- Peptide mapping with and without PNGase F treatment (an enzyme which catalyzes hydrolysis of N-linked glycans from the protein backbone) to identify and determine the occupancy level of the *N*-linked site(s).
- Glycan mapping for *N*-glycan composition analysis.
- Structural characterization by sequential MS of glycan fractions (may be permethylated to facilitate analysis) and exoglycosidase analysis.

In addition to elucidating the major glycoforms, the presence of non-human glycan moieties also may be assessed given that they may be of potential immunogenic concern (e.g., glycans bearing *N*-glycolylneuraminic acid, a non-human sialic acid, and galactose-α-1,3-galactose) (*18*, *19*).

To investigate the biological effect of aglycosylated variants, the product can be treated with PNGase F, purified, and tested for potency relative to a control sample without PNGase F (but otherwise treated identically).

Disulfide Structure

Proteins containing cysteine residues may display disulfide-mediated structural isoforms, including unpaired cysteines (e.g., free sulfhydryls), and cysteines with various pairing configurations if more than two cysteines are present and the structural configurations allow for isoform formation. For example, multiple disulfide structural isoforms are inherent to recombinant and naturally occurring IgG2 molecules (*20*). The connectivity of disulfide bonds may be elucidated using non-reduced and reduced peptide maps coupled with electrospray ionization-MS (ESI-MS) for identification. Free sulfhydryls are often present and can be quantified by an Ellman's or similar test (*21*). Contemporary literature includes reports on the presence of other cysteine-mediated bonding in recombinant proteins, including trisulfide (*22*) and thioether (*23*) linkages. These variants differ by the number of sulfur atoms in the linkage, and therefore can be detected by MS as a mass change corresponding to the presence of an additional sulfur (trisulfide) or the loss of a sulfur (thioether).

Determination of the biological activity of disulfide isoforms can be challenging unless there is a separation mode that resolves or sufficiently enriches them under native conditions (e.g., ion-exchange or size exclusion chromatography.) In some cases, manipulating redox potential through redox agents or modifying pH can enrich certain isoforms for potency measurements.

Charge Variants

Proteins commonly display several sources of charge heterogeneity arising from post-translational modifications of the desired product. Typical modifications that contribute to a complex charge profile representing a mixture of product-related variants and/or product-related impurities may include, but are not limited to, C-terminal heterogeneity (*24*), N-terminal pyroglutamate formation (*1*), deamidation, and oxidation (e.g., methionine oxidation can impact the surface charge by modifying the local pK_a of ionizable side chains through subtle structural environment changes).

Charge heterogeneity of proteins is typically evaluated by ion exchange-high-performance liquid chromatography (IEX-HPLC) and capillary isoelectric focusing (cIEF) (Separation chapter/Volume 2, Chapter 5). IEX-HPLC separates proteins based primarily on the heterogeneity of surface charge; however, it also may be influenced by structural heterogeneity and other modifications that affect molecular interactions with the ion exchange resin. Consequently, the observed chromatographic profile may result from a combination of charge differences and structural variants of proteins.

Characterization of acidic and basic peak fractions can be accomplished by fraction collection followed by peptide mapping and other orthogonal analytical techniques. The potency of these fractions also can be determined. When fractionating peaks for further analysis, a control sample also should be prepared (e.g., by collecting the main peak fraction or combining all fractions into a control sample to factor in the sample handling and solution conditions that can result in experimental artifacts.)

cIEF is an orthogonal technique to assess the charge heterogeneity of a protein. Charge variants are separated by differences in their isoelectric point (pI) and are separated as they migrate within a pH gradient established during the focusing step under the influence of an applied electric field. The pI of acidic, main, and basic peak regions are determined based on known pI markers. Although the capillary format of the technique precludes isolation and further analysis of individual fractions directly from the cIEF profile, correlative strategies using preparative IEF can aid in the identification of charge variants.

Size Variants and Heterogeneity by Size Exclusion Chromatography (SEC)

Size heterogeneity under native solution conditions typically is monitored by non-denaturing SEC, which generally resolves submicron aggregates (i.e., HMW species) from the monomer in the main peak (Separation chapter/Volume 2, Chapter 5). Low molecular weight (LMW) species are typically poorly resolved and/or co-elute with excipient peaks, and are better monitored by reduced or non-reduced capillary electrophoresis with sodium dodecylsulfate (rcSDS or nrcSDS). Examples of common protein size modifications that would be typically detected by SEC are:

- Aggregation through noncovalent interactions, such as hydrophobic interactions and/or salt bridges.
- Covalent modifications that form intermolecular covalent linkages, such as disulfide cross-linking.

Monomer and submicron aggregates can be purified by fraction collection of peaks from the SEC method, and subsequent analyses such as cSDS and rcSDS can help elucidate the covalent nature of the submicron aggregates. Potency measurements can also be performed on the fractions to assess any potential impact of the submicron aggregates on efficacy.

Addition of static light scattering (SLS) detection to the SEC method allows the determination of molar mass for individual peaks in the chromatogram.

The intensity of light scattered by an eluting species is proportional to both the concentration and molecular weight of the species. The intensity of UV absorbance (280 nm) is proportional to protein concentration. The molar mass of each eluting species can be determined by the instrument manufacturer's software by utilizing the light scattering intensity and concentration for each peak.

Size Heterogeneity by SV-AUC

Sedimentation velocity analytical ultracentrifugation (SV-AUC) is a characterization method that provides size and conformation information directly from a sample in solution (Biophysical chapter/Volume 2, Chapter 6). This can be utilized as a complementary technique to fractionation-based techniques such as SEC to confirm there is no perturbation of weak self-association equilibria or exclusion of higher order oligomers from the chromatographic separation and as such can be used to qualify or validate SEC method accuracy. There is minimal matrix interaction and sample handling during SV-AUC analysis. While SV-AUC has many advantages over SEC, the main drawbacks are a relatively high limit of quantitation and a long and complex analytical procedure that limits throughput. The SV-AUC method is typically developed and qualified to provide optimal resolution and precision for the detection of submicron aggregates. The continuous distribution is the result of model fitting in which a spectrum of species, each having a different sedimentation coefficient (S value), is used to describe the raw sedimentation data. The sedimentation coefficient of each species is a function of molecular size and shape. The relative concentration (i.e., weighting) of each species is shown on the y axis of the continuous distribution. Each peak in the continuous distribution can be integrated, and its area (as a percentage of the total area) represents the relative concentration of that species.

Size Heterogeneity by cSDS

LMW and non-reducible submicron aggregate-size heterogeneity in disulfide-containing proteins can be obtained by reducing all of the disulfide bonds and separating the constituent units using rcSDS (Separation chapter/Volume 2, Chapter 5). Examples of common protein size modifications that would be typically detected by rcSDS are:

- Hydrolysis that cleaves the polypeptide chains.
- Non-reducible covalent linkages.

cSDS can also be performed under non-reducing conditions to evaluate the presence of covalent (disulfide or non-disulfide linked) submicron aggregate forms, as well as LMW forms in both disulfide- and non-disulfide-containing proteins. Fractionation of product variants by cSDS for further characterization (e.g., peptide mapping, bioassay) is not practical. However, denaturing (and reduced-denaturing) size-exclusion chromatography methods, although not as resolving as the cSDS technique, may be employed to enrich some of the size variants for further analysis. Any denaturing technique will likely impact potency due to perturbation of the native structure, and therefore, potency measurements may not add value in the characterization of size variants isolated by denaturing techniques. However, submicron aggregates and LMW species (with the exception of dimers in some cases) are typically categorized as CQAs and are routinely incorporated as part of the product control strategy.

Size Heterogeneity by dSEC

Similar to cSDS methods, denaturing SEC (under reduced and non-reduced conditions) provides an understanding of the covalent and noncovalent bonding present in monomer and size variants. For example, if a size variant is present under non-denaturing conditions but is dissociated to monomer subunits by denaturants, then it can be concluded that hydrophobic interactions or salt bridges bond the monomer subunits together to form the size variant. If the size variant persists under denaturing conditions but dissociates under reducing conditions, then it can be concluded that its structure is bound through disulfide bonds. If the size variant persists under denaturing and reducing conditions, it must be covalently linked by non-disulfide bond(s).

Biophysical Characterization

The primary amino acid sequence of a protein contributes to the final secondary and tertiary structural conformation(s). Evaluation of biophysical properties such as photo-absorption, photo-emission, and thermal stability provides information about these higher order structures and overall protein conformation (Biophysical chapter/Volume 2, Chapter 6). Assessment of these properties can include a comparison to similar protein structures (or class of structures, as in IgG1 and IgG2 structural classes) to assess if the overall structure is consistent with that expected from similar molecules.

Secondary Structure

Analysis of secondary structure examines the inter-residue interactions mediated by hydrogen bonds. The most common structural motifs of protein secondary structure are α-helices and β-sheets. The secondary structure of predominantly β-sheet containing proteins are usually elucidated using Fourier transform infrared (FTIR) spectroscopy, predominantly by measuring absorption profiles in the amide I region and deconvoluting with second derivative analysis to estimate secondary structure content. The secondary structure of predominantly α-helix containing proteins are usually elucidated using far-UV circular dichroism (far-UV CD) spectroscopy, where the mean residue ellipticity is assessed as a function of wavelength. Amide chromophores absorb between approximately 190–250 nm, with profiles determined predominantly by hydrogen bonding structures that are in turn driven by specific secondary structures. The spectral profiles can be plotted and predominant secondary structures are either estimated based on the spectral profile or deconvoluted with mathematical algorithms to estimate secondary structure content. Precise secondary structure determinations are generally not needed, the goal mainly being to determine the major secondary structure(s) elements.

Tertiary Structure

The tertiary structure of a protein is governed by various interactions among different regions of the molecule, including hydrogen bonds, hydrophobic interactions, and salt bridges. The tertiary structure of proteins is commonly elucidated using near-UV CD spectroscopy, where the mean residue ellipticity is assessed as a function of wavelength. Aromatic residues absorb between approximately 250–350 nm, with profiles determined predominantly by the local environment of the aromatic residue (solvent-exposed or buried within the protein structure).

Thermal Stability

The thermal stability of a protein (or specific region of a protein) is regulated by the manner in which the protein is folded; therefore, assessment of this characteristic reveals information on the protein's conformation. The thermal stability is typically assessed using differential scanning calorimetry (DSC), which measures excess heat capacity as a function of temperature. As temperature is increased, proteins unfold at characteristic thermal melting temperatures that are dependent on the thermodynamic stability of the folded protein (or folded protein domains.)

Forced Degradation Pathways

Understanding of the mechanisms by which a therapeutic molecule responds to specific stress conditions is recommended per ICH Q6B, Q5E, Q1A(R2) and Q5C (*2*, *4*, *6*, *7*). Stress studies provide insight into a number of important aspects of the product, including identification of potentially relevant PQAs, elucidation of potential stress conditions for comparability assessments, and demonstration of stability-indicating properties of methods, thereby providing orthogonal confirmation of product variants resolved by the purity methods (e.g., correlating submicron aggregates levels in stressed samples by SV-AUC and SEC) that in turn provides insight into potential product impact (and root cause) if product is exposed to an excursion from typical manufacturing or sample handling conditions.

There are many degradation conditions available for understanding the susceptibility and degradation pathways of the product under stressed conditions. Some of these conditions are summarized in Table 2 and include aggregation, LMW species (e.g., fragmentation, clips), and typical biochemical modifications of specific residues (e.g., methionine oxidation, asparagine deamidation) in response to the stressed condition applied.

Although stress conditions can readily foster product degradation, commercial manufacturing, storage, and transport conditions are controlled to minimize product degradation. As such, the identified degradation pathways presented in Table 2 would be expected to occur at significantly lower rates (if at all) during manufacturing, storage, and/or transportation than the rates observed in these forced degradation studies.

Table 2. Potential Forced Degradation Conditions and Degradation Products

Stressed Condition	Typically Observed Degradation Products	Example Dominant Degradation Pathway
Thermal exposure (50°C)	Submicron aggregates, LMW species	Aggregation
Low pH (3.5)	Submicron aggregates, LMW species	Aggregation, fragmentation
Physiological/high pH (7.4)	Deamidation, submicron aggregates, LMW species	Deamidation
Chemical oxidation	Methionine oxidation	Oxidation
Light exposure	Methionine oxidation, submicron aggregates, LMW species	Oxidation, including Trp oxidation
Agitation	HMW, subvisible, and visible particles	Product-dependent

The formation of degradation products induced by each stress condition over time may be monitored by relevant product quality and characterization methods, including the assessment of size variants (SEC and rcSDS), charge variants (cation exchange-HPLC [CEX-HPLC]), biochemical modification (peptide mapping), higher order structure (as needed), and biological properties (potency and receptor-binding methods). For each stress condition, samples are typically removed at time points over the duration of the study and frozen. After the final time point, all samples are simultaneously submitted for analysis by each method to minimize analytical variability.

Product stability is continually evaluated at multiple, well-controlled conditions as part of the GMP stability programs. Degradation pathways under forced degradation conditions may be explored in order to ensure controls are established, appropriate conditions are utilized to maintain the stability of the product, and appropriate methods are developed and incorporated into the existing stability program.

Biological Characterization

Comprehensive biological characterization of the product is typically conducted to establish the relevance of the biological assay(s) to the product's MOA and assessment of specificity of relevant binding domain(s), if applicable. In addition, biological characterization of the product may include an assessment of the structure-function properties of variants, as well as their classification as product-related substances or product-related impurities, thereby contributing to the assessment of attribute criticality.

Overall Conclusions on Elucidation of Structure

The structure of protein products may be elucidated from a variety of biological, biochemical, biophysical, and forced degradation and/or stability techniques to provide a comprehensive understanding of the structure-functional properties of the bulk drug substance, assessment of product-related variants, and associated attribute criticality determination.

The integrity of the intact protein can be confirmed by whole mass analysis, and the complete amino acid sequence can be confirmed to the extent possible (and post-translational modifications detected) through MS of the intact protein combined with proteolytic digestion using complementary endoproteases and comparison of peptide fragmentation patterns to predicted fragmentation patterns.

Comprehensive characterization of the glycan structures can be obtained using separation techniques coupled with MS with exoglycosidase treatments and/or labeling techniques, and results can be compared with a composition typical of mammalian *N*-linked glycans, with attention paid to the identification of any potential immunogenic glycans.

The disulfide structure can be determined using comparative peptide mapping under non-reducing and reducing conditions. Free sulfhydryl content may be probed under native and denaturing conditions with Ellman's assay (or comparable reagent-based techniques).

Charge heterogeneity is often evaluated by IEX-HPLC and cIEF methods. Charge variants can be purified by IEX-HPLC and subjected to further biochemical and biological characterization to assess attribute criticality.

Submicron aggregate product attributes are often assessed by SEC under non-denaturing and denaturing conditions (which enables further purification and characterization) and by cSDS (which can also assess LMW species). Submicron aggregates and LMW species are typically considered as high criticality quality attributes, since they don't have the structure of the intended product and have immunogenic potential.

The biophysical properties of the protein are typically assessed by FTIR spectroscopy, near-UV CD spectroscopy, and DSC to ascertain the secondary and tertiary structural conformations and relative thermodynamic stability. The biophysical evaluation helps establish that the protein is properly folded and contains well-defined secondary and tertiary structures that are consistent with other similar motifs (as applicable).

Relevant Degradation Pathways under Recommended Storage Conditions

Protein stability is critical for successful use of biopharmaceutical products. Many antibody-based therapeutics have expiry periods up to 2–3 years; they must remain stable for the duration of the expiry period to be useful as medicines. Biopharmaceutical firms expend significant resources on formulation development to maximize protein stability over time, and most biopharmaceutical products have recommended storage conditions that include refrigeration (i.e., 2–8°C). Protein degradation falls into two major categories: physical and chemical. It is important to note that biopharmaceutical products possess different biochemical attributes, are formulated in different buffer systems, can be contained within different dosage forms (i.e., vials vs. syringes), and can be either liquids or solids (lyophilized product). Thus, degradation vulnerabilities and pathways will differ between products, and stability needs to be evaluated on a product-specific basis.

Chemical Degradation

Important chemical degradation pathways include the following, although this list is not all-inclusive:

- Deamidation. This involves a side chain of an asparagine reacting with the following peptide group, resulting in either an aspartate or an isoaspartate. Because this changes the overall charge of a protein, it can be detected by methods that evaluate charge-based changes such as cIEF, discussed later in the Separation chapter/Volume 2, Chapter 5.
- Oxidation. Oxidation of methionine residues to methionine sulfoxide can occur during cell culture or if peroxides are present in formulation buffers. Because these changes result in only modest molecular weight changes, they do not result in charge shifts and often occur in only a small percentage of the protein product; methods such as tryptic peptide mapping is often required to detect and quantify oxidation.
- Clipping. Slow protein clipping can occur if residual enzymes from the host cell system are still present after purification. In certain drug products, such as prefilled syringes, metals leaching from the container system can catalyze clipping as well. If significant levels of clipping occurs over time, it can be detected as the accumulation of LMW fragments detectable in SDS-polyacrylamide gel electrophoresis (SDS-PAGE) or related methods.

Physical Methods

From a biopharmaceutical standpoint, aggregation and particle formation, described below, should be strongly considered for inclusion in a stability program.

- Aggregation. Aggregation is the formation of complexes of proteins over time in solution or upon resuspension, in the case of lyophilized products. Generally, the term "aggregation" is considered to apply to complexes small enough to be analyzed by such methods as SEC. These include dimers, trimers and other multimers but not complexes large enough to be visible. Aggregation is generally believed to be nucleated by low-level protein unfolding and can be covalent or noncovalent. Aside from SEC, other methods described in this book that can measure aggregates include FFF, AUC, and several others.
- Particles. Particles are complexes that are larger than aggregates and probably cannot be detected by SEC because they would get trapped in the column. Particle formation occurs when aggregates grow in size over time, often to the point where they are visible. Container closure issues can contribute to particle formation, for example, by contributing low levels of silicone oil to nucleate their formation over time. They generally fall into two categories: micron-scale aggregates (i.e., subvisible particles), which can be measured by methods such as light obscuration; and aggregates greater than 100 microns (i.e., visible particles), where the human eye is often the most sensitive test.

PQA Functional Assessment

With the advent of advanced analytical methods such as those described in this book, it is clear that protein biopharmaceuticals possess significant and subtle variability both between and within batches. When viewed from a combinatorial basis, a single batch of an antibody, in theory, could contain thousands of individual variants at various levels. The questions that this raises are which of the product variants are significant and should be evaluated, and whether they can be identified from elucidation of structure. The significance of some variants is obvious; for example, clipping between the variable and Fc part of an antibody will ablate effector functions if present at high levels. Others are less so; for example, would deamidation at a residue far from either the complementarity-determining region (CDR) surface or an identified effector part of the Fc part of antibody impact function? In between these two extremes are a broad range of borderline cases for which an argument could be made in either direction: significant or not significant.

In cases such as these, utilization of platform, literature, and other prior knowledge (e.g., IgG1 and IgG2 mAb platforms) is a key part of an initial risk assessment of variants, especially for Fc-related variants. For some variants, especially those at very low levels, a knowledge space-based assessment may be initially sufficient. Borderline cases may require an experimentally based approach to pin down the biological relevance of variants, especially major species. The classic example of this is fucosylation of Fc N-linked glycans, which are known to impact antibody-dependent cell cytotoxicity (ADCC) (25). This attribute is probably important for the *in vivo* function of most but not all antibody-based therapeutics. The risk-based assessment and identification of CQAs in the context of what is known to be the *in vivo* function of the protein product is a hallmark of the QbD approach for biopharmaceutical development. Although the details of a QbD assessment will by necessity vary from product to product, the knowledge space developed for previous variations of closely related classes of products such as antibodies is always germane to consider when developing a new biopharmaceutical product.

Acknowledgments

The authors thank Scott Lute, Cyrus Agarabi and Michael Boyne (Office of Pharmaceutical Science, Center for Drug Evaluation and Research) for their careful review of this manuscript.

References

1. *Guidance for Industry Q8 Pharmaceutical Development.* International Conference on Harmonisation (ICH) of Technical Requirements for Registration of Pharmaceuticals for Human Use, Geneva, Switzerland, 2005.
2. *Guidance for Industry Q6B Specifications: Test Procedures and Acceptance Criteria for Biotechnological/Biological Products.* International Conference on Harmonisation (ICH) of Technical Requirements for Registration of Pharmaceuticals for Human Use, Geneva, Switzerland, 1999.
3. *Guidance for Industry Q9 Quality Risk Management.* International Conference on Harmonisation (ICH) of Technical Requirements for Registration of Pharmaceuticals for Human Use, Geneva, Switzerland, 2005.
4. *Guidance for Industry Q5E Comparability of Biotechnological/Biological Products Subject to Changes in Their Manufacturing Process.* International Conference on Harmonisation (ICH) of Technical Requirements for Registration of Pharmaceuticals for Human Use, Geneva, Switzerland, 2004.
5. *Guidance for Industry Q2(R1) Validation of Analytical Procedures: Text and Methodology.* International Conference on Harmonisation (ICH) of Technical Requirements for Registration of Pharmaceuticals for Human Use, Geneva, Switzerland, 2005.
6. *Guidance for Industry Q1A(R2) Stability Testing of New Drug Substances and Products.* International Conference on Harmonisation (ICH) of Technical Requirements for Registration of Pharmaceuticals for Human Use, Geneva, Switzerland, 2003.
7. *Guidance for Industry Q5C Quality of Biotechnological Products: Stability Testing of Biotechnological/Biological Products.* International Conference on Harmonisation (ICH) of Technical Requirements for Registration of Pharmaceuticals for Human Use, Geneva, Switzerland, 1996.
8. Chelius, D.; Jing, K.; Lueras, A.; Rehder, D. S.; Dillon, T. M.; Vizel, A.; Rajan, R. S.; Li, T.; Treuheit, M. J.; Bondarenko, P. V. *Anal. Chem.* **2006**, *78*, 2370–2376.
9. Bradshaw, R. A.; Brickey, W. W.; Walker, K. W. *Trends Biochem. Sci.* **1998**, *23*, 263–267.

10. Antes, B.; Amon, S.; Rizzi, A.; Wiederkum, S.; Kainer, M.; Szolar, O.; Fido, M.; Kircheis, R.; Nechansky, A. *J. Chromatogr. B.* **2007**, *852*, 250–256.
11. Walsh, G.; Jefferis, R. *Nat. Biotechnol.* **2006**, *24*, 1241–1252.
12. Johnson, K. A.; Paisley-Flango, K.; Tangarone, B. S.; Porter, T. J.; Rouse, J. C. *Anal. Biochem.* **2007**, *360*, 75–83.
13. Ji, J. A.; Zhang, B.; Cheng, W.; Wang, Y. J. *J. Pharm. Sci.* **2009**, *98*, 4485–4500.
14. Harmon, P. A.; Kosuda, K.; Nelson, E.; Mowery, M.; Reed, R. A. *J. Pharm. Sci.* **2006**, *95*, 2014–2028.
15. Flynn, G. C.; Chen, X.; Liu, Y. D.; Shah, B.; Zhang, Z. *Mol. Immunol.* **2010**, *47*, 2074–2082.
16. Prien, J. M.; Prater, B. D.; Cockrill, S. L. *Glycobiology* **2010**, *20*, 629–647.
17. Mechref, Y.; Hu, Y.; Desantos-Garcia, J. L.; Hussein, A.; Tang, H. *Mol. Cell. Proteomics* **2013**, *12*, 874–884.
18. Chung, C. H.; Mirakhur, B.; Chan, E.; Le, Q. T.; Berlin, J.; Morse, M.; Murphy, B. A.; Satinover, S. M.; Hosen, J.; Mauro, D.; Slebos, R. J.; Zhou, Q.; Gold, D.; Hatley, T.; Hicklin, D. J.; Platts-Mills, T. A. *N. Engl. J. Med.* **2008**, *358*, 1109–1117.
19. Ghaderi, D.; Taylor, R. E.; Padler-Karavani, V.; Diaz, S.; Varki, A. *Nat. Biotechnol.* **2010**, *28*, 863–869.
20. Wypych, J.; Li, M.; Guo, A.; Zhang, Z.; Martinez, T.; Allen, M.; Fodor, S.; Kelner, D.; Flynn, G. C.; Liu, Y. D.; Bondarenko, P. V.; Speed-Ricci, M.; Dillon, T. M.; Balland, A. *J. Biol. Chem.* **2008**, *283*, 16194–16205.
21. Zhang, W.; Czupryn, M. J. *Biotechnol. Prog.* **2002**, *18*, 509–513.
22. Pristatsky, P.; Cohen, S. L.; Krantz, D.; Acevedo, J.; Ionescu, R.; Vlasak, J. *Anal. Chem.* **2009**, *81*, 6148–6155.
23. Tous, G. I.; Wei, Z.; Feng, J.; Bilbulian, S.; Bowen, S.; Smith, J.; Strouse, R.; McGeehan, P.; Casas-Finet, J.; Schenerman, M. A. *Anal Chem.* **2005**, *77*, 2675–2682.
24. Harris, R. J. *J. Chromatogr. A.* **1995**, *705*, 129–134.
25. Shinkawa, T.; Nakamura, K.; Yamane, N.; Shoji-Hosaka, E; Kanda, Y.; Sakurada, M.; Uchida, K.; Anazawa, H.; Satoh, M.; Yamasaki, M.; Hanai, N.; Shitara, K. *J. Biol. Chem.* **2003**, *278*, 3466–73.

Chapter 5

Using Quality by Design Principles in Setting a Control Strategy for Product Quality Attributes

Gregory C. Flynn[*,1] and Gregg B. Nyberg[2]

[1]Process and Product Development, Amgen Inc.,
Thousand Oaks, California 91320, United States
[2]Drug Substance and Analytical Technologies, Amgen Inc.,
Longmont, Colorado 80503, United States
*E-mail: gflynn@amgen.com

Utilizing quality by design (QbD) principles in the development and manufacturing of a biotherapeutic helps to ensure a safe and efficacious product. A manufacturing process should be designed to consistently produce product with the desired quality attributes. Risk assessment tools are needed to systematically evaluate the capacity of a developed process to deliver product with acceptable quality. The Product Quality Risk Assessment (PQRA) described in this chapter utilizes a failure modes and effects (FMEA) approach that considers severity, likelihood of occurrence, and detection to determine an overall risk level. This assessment begins by evaluating the product quality attributes (PQAs) and scoring their relative criticality (severity). Next, the knowledge of the relationship between process and PQA level is used to determine the likelihood of occurrence that individual process steps will cause a quality attribute level to deviate outside of its acceptable range. The attribute severity and likelihood of occurrence define a preliminary hazard level that can be used to prioritize process and product development activities on areas of highest risk. Finally, the ability to detect quality attribute deviations and prevent their impact to patients is taken into account to establish the overall risk level. Two examples (Fc high-mannose glycans [HMGs] and glycation) are provided to demonstrate how a

PQRA would be conducted in practice, from studies designed to probe for the biological impact of a PQA to the final risk assessment.

Introduction

Quality by design (QbD) is defined by the International Conference on Harmonisation as a systematic approach to development that begins with predefined objectives and emphasizes product and process understanding and process control, based on sound science and quality risk management (*1*). Comprehensive understanding of the disease state and how the drug product's molecular attributes impact the biological mechanism(s) of action and safety of the molecule is foundational to QbD. The manufacturing process should be designed to consistently achieve product with the desired quality attributes. Comprehensive understanding of how the process impacts product quality attributes (PQAs) is required to ensure effective process design and develop an effective control strategy. The term "control strategy" refers to the combination of input, procedural, and testing controls applied to ensure a process that consistently delivers product meeting PQA requirements (see Figure 1).

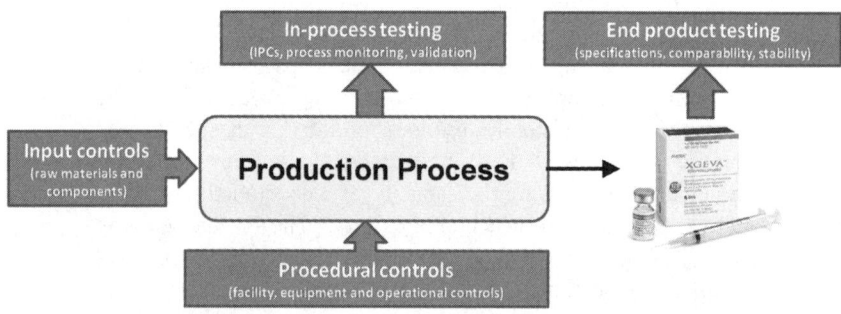

Figure 1. Control strategy overview.

The control strategy comprises a number of individual control elements that are applied as required to ensure adequate control. Table 1 summarizes various control elements and where they are described in the Common Technical Document (CTD) for regulatory filings. The level of control for each individual quality attribute is determined on the basis of the criticality level of the attribute to safety and efficacy and the capability of the process to consistently deliver product that meets acceptance criteria. It is the sum of the individual control strategies for each quality attribute that represents the overall process control strategy. Using risk assessment as a tool, a rational QbD control strategy can be formulated for each quality attribute by choosing the appropriate control elements.

Table 1. Description of Control Elements[a]

Control Element	Description
Procedural	Facility, equipment, and/or operational control that ensures robust and reproducible operations supporting control of a specific quality attribute, including operational limits placed on critical process parameters (3.2.S.2.2 and 3.2.P.3.3, Description of Manufacturing Process and Process Controls)
Raw material	Controls pertaining to raw materials, excipients, components, etc. used in manufacturing operations that serve as a direct control for a specific quality attribute (3.2.S.2.3, Raw Materials, and 3.2.P.4, Control of Excipients)
Clinical lot evaluation	Testing and/or evaluation performed during manufacture of historical clinical lots (3.2.S.4.4 and 3.2.P.5.4, Batch Analyses)
Comparability	Testing performed to demonstrate comparable attributes following significant process changes, including process and product comparability (3.2.S.2.6, Comparability)
Characterization	Testing performed once or infrequently on representative material to enhance process and product understanding (3.2.S.2.6, Process Characterization; 3.2.S.3.1, Elucidation of Structure and Other Characteristics; 3.2.S.3.1, Biological Characterization; 3.2.P.3, Pharmaceutical Development)
Process performance qualification (PPQ)	Testing performed during PPQ lots; used as a means to confirm process design and demonstrate the commercial manufacturing process performs as expected (3.2.S.2.5 and 3.2.P.3.5, Process Validation and Evaluation)
In-process controls	Testing performed on an every-lot basis to ensure that the drug substance and drug product conform to specifications; in-process control parameters have action and/or rejection limits assessed as part of lot disposition (3.2.S.2.4 and 3.2.P.3.4, Control of Critical Steps and Intermediates)
Process monitoring	Statistical evaluation of parameters used to identify shifts in performance and trends; used as a means to provide continual assurance that the process remains in a state of control during commercial manufacture (3.2.S.2.4 and 3.2.S.3.4, Control of Critical Steps and Intermediates)
Specification	Tests with associated acceptance criteria conducted at final lot release to confirm quality of the drug substance or drug product (3.2.S.4.5 and 3.2.P.5.6, Justification of Specifications)

[a] Control elements cited refer to the Common Technical Document for regulatory filings (www.ich.org/products/ctd.html).

A systematic approach to performing QbD Product Quality Risk Assessment (PQRA) is described in this chapter. An overview of the approach is shown in Figure 2. A list of PQAs is compiled and evaluated in a PQA Assessment. Because the term "critical quality attribute" refers to a chemical or physical modification that must be controlled within acceptable ranges or limits to achieve the desired product quality (*1*); a different term is needed for those attributes prior to such judgments about control. For this we use the term "product quality attribute." The output of the PQA Assessment is a list of PQAs scored for relative criticality. The PQRA combines the PQA criticality with knowledge of the process capability and the testing strategy to determine the overall risk to the patient. The tool described can be applied at various stages of drug development to ensure process design activities are prioritized to address areas of highest risk. This chapter will use two antibody PQA case study examples to demonstrate how the tool can be used to develop effective control strategies.

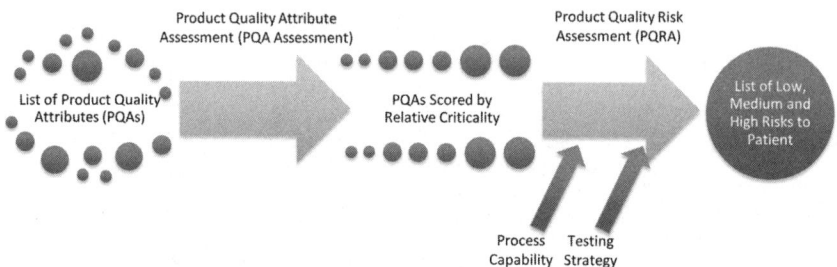

Figure 2. Overview of a systematic approach to Product Quality Risk Assessment (PQRA).

Introduction to Product Quality Risk Assessment (PQRA)

The PQRA described in this chapter utilizes a failure modes and effects (FMEA) approach that considers severity, likelihood of occurrence, and detection to determine an overall risk level (Figure 3). This assessment begins by evaluating the PQAs and scoring their relative criticality (severity). Next, the knowledge of the relationship between process and PQA level is used to determine the likelihood of occurrence that individual process steps will cause a quality attribute level to deviate outside of its acceptable range. The attribute criticality (severity) and likelihood of occurrence define a preliminary hazard level that can be used to prioritize process and product development activities on areas of highest risk. Finally, the ability to detect PQA deviations and prevent their impact to patients is taken into account to establish the overall risk level.

The output from the PQRA is the risk associated with a defined process and control strategy. The PQRA is designed to be used in an iterative fashion to refine the control strategy to achieve a desired risk profile. The control strategy could be refined through changes to process capability (likelihood of occurrence), product understanding (severity), method capability (detection), or testing strategy (detection).

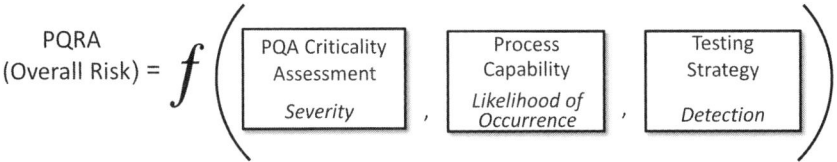

Figure 3. Product quality risk assessment (PQRA) approach to determine overall risk to the patient.

Product Quality Attribute (PQA) Assessment

The first step in the PQRA process is to determine the severity score for each PQA. This can be obtained through a separate assessment, called here the "Product Quality Attribute Assessment." Examples provided here will be restricted to antibodies, but the general approach is valid for any therapeutic protein. In this approach, each therapeutic antibody PQA is evaluated or ranked based on its impact on safety and efficacy. Because the PQRA likelihood-of-occurrence score (process capability) pertains to the process' ability to maintain or limit the level of an attribute, the range or level of the attribute found in the process should not be considered in the PQA Assessment. Instead, the evaluation considers the impact if the attribute were not controlled (i.e., elevated or reduced compared to expected levels). The output of the PQA Assessment is a severity score ranging from a severe effect (causing significant harm to a patient) to an insignificant effect (no impact is expected). In the example used here, the criticality is given a numerical score from 1 (least severe) to 9 (most severe).

PQAs may impact the drug's safety or efficacy. A PQA has the potential to generate a new epitope on the drug, or the level of the PQA may increase the rate of anti-drug antibody production. Other non-immune safety issues also may be affected by PQA levels. Off-target activities may be enhanced by a particular attribute or may lead to a toxic effect by an unknown mechanism. PQAs also may have multiple effects on efficacy. Target binding, effector function activities, and pharmacokinetics (PK) all have the potential to be affected by PQAs. Each of these safety and efficacy effects should be considered for each attribute. One way to ensure proper evaluation is to score each potential impact as a separate subscore such as immune safety, non-immune safety, potency, and PK. For each scoring category, the attribute is ranked for patient impact. Tables 2 and 3 provide scoring guidance on how a PQA could be scored for each subcategory. Examples of attribute scoring are shown for two attributes, Fc high-mannose glycans (HMGs) and glycation, discussed later.

PQAs refer to physical or chemical characteristics of the molecule, not the methods used to detect these characteristics. In this way, the impact of the attribute on safety and efficacy can be judged independent of a particular method to detect and quantify that attribute. Increasing the specificity of a PQA assay can often enhance the assay's value. For example, the effect of a deamidation event, such as on Asn 384 in the Fc, can be judged for its biological impact on a specific antibody or on antibodies in general (2). Although the deamidation at Asn 384 exists at a certain level, the ability of each method to detect and accurately quantify that deamidation will vary. Peptide mapping is considered a very selective method that has the potential to quantify deamidation at one position independent of others. Cation exchange chromatography (CEX), in contrast, may be less selective. Deamidation at Asn 384 may result in the change in the acidic peak area, but other species may also exist within that peak. Therefore, the acidic peak should not be considered a PQA. Whether to use CEX to measure Asn 384 deamidation or employ a more selective method would be considered during the PQRA.

Table 2. Product Quality Attribute (PQA) Severity Scoring for Safety Subscores

Severity Score	Immunogenicity (IG) Potential for:	Non-Immune Safety Potential for:
Severe (9)	• PQA may enhance IG AND • Clinical experience: High IG and hypersensitivity or loss of endogenous protein function observed	• Death • Serious patient injury • Microbiologically related infections
Major (7)	• PQA may enhance IG AND • Clinical experience: High IG with change in therapeutic function or unknown IG in patients with increased immune activity	• Hospitalization • Toxicity known to be associated with PQA• Undesirable change in PQA after pivotal trials
Moderate (5)	• PQA may enhance IG AND • Clinical experience: High IG with no observed impact, low IG with change in therapeutic function or unknown IG in patients with normal or low immune activity	• Treatment required without hospitalization • Impact to patient • Undesirable change in PQA after Phase 2 trials • Moderate to high concentration of PQA (greater than level qualified in toxicology studies)
Minor (3)	• PQA may enhance IG AND • Clinical experience: Low IG with no observed impact	• Adverse event not requiring treatment • Customer annoyance • Chronic treatment or disease • Cell-based therapeutic target
Insignificant (1)	• PQA not expected to enhance IG	• No known safety issues • Acute treatment only • Serious disease with no alternative treatments

Table 3. Product Quality Attribute (PQA) Severity Scoring for Efficacy Subscores

Severity Score	*Pharmacokinetic (PK)* *Potential for:*	*Potency* *Potential for:*
Severe (9)	Progression of disease due to change in PK	Progression of disease due to change in potency or effector function
Major (7)	Major change in PK and/or pharmacodynamics (PD)	Major change in potency or effector function
Moderate (5)	Moderate change in PK and/or PD	Moderate change in potency or effector function
Minor (3)	Minor change in PK and/or PD	Minor change in potency or effector function
Insignificant (1)	No expected change in PK and/or PD	No expected change in potency or effector function

Only PQAs thought to exist under reasonable conditions should be scored. These conditions would include excursions in the process or storage but would not include harsh artificial chemical or physical stresses on the therapeutic antibody that it would not reasonably encounter. Early in development, however, the analytical characterization information may be limited, so there may be insufficient knowledge of a PQA's existence. In this case, the assumption should be made that the attribute exists and should be scored until proven otherwise. Before deciding not to score an attribute, subject matter experts should consider the capabilities of the analytical technique to detect and quantify the PQA in question.

A variety of sources and techniques can be applied to help assess an attribute's biological impact. The confidence placed on specific data will depend on its relevancy to the particular therapeutic antibody, its mechanism of action (MoA), and general data quality. PQAs with known safety or efficacy impacts would generally be considered a greater concern than ones with no or questionable impact information. New information on PQA impacts that is gathered as a drug progresses through development would be used to revise the severity scoring.

One useful technique that can be applied to better understand a variety of attributes is the study of the PQAs *in vivo* (*3*). An attribute may change over time while in circulation either by affecting clearance or by converting into other forms. Samples from animal PK studies done prior to clinical trials or early PK studies in humans can be used to support both safety and efficacy scores. If an attribute affects clearance, the impact can be estimated (see the example in Appendix A). The degree to which clearance is affected and the significance of clearance changes on drug efficacy are both considered when adjusting the scoring. Attribute conversion when combined with PK can predict patient exposure to the attribute. Understanding how the combination of lot-to-lot variability and conversion *in vivo* affects overall patient exposure will help in setting reasonable ranges and specifications for the PQA.

Figure 4 illustrates the experimental outline and results from a PQA *in vivo* study. An antibody drug heterogeneous for a particular PQA, designated as a triangle and square in the figure, is injected into humans as part of a PK study. The combination of these two forms together sum to 100% of the product. For example, these two symbols could represent the deamidated and unmodified forms on one primary site on the molecule. Product quality from the serum samples withdrawn from patients over time can be determined, providing the changes in the relative levels of these attributes with time (Figure 4, upper right). When the results from the attribute changes are combined with the PK results (Figure 4, upper central), a plot can be generated indicating the changes in the drug containing the two forms over time. Examples of how this information can be used are provided in the two boxes.

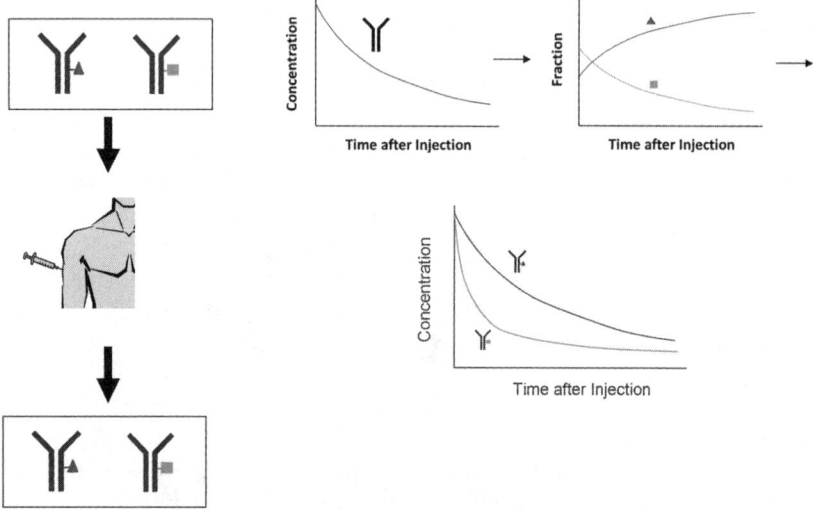

Figure 4. Product quality attribute changes in vivo study.

Likelihood of Occurrence (Process Capability)

Likelihood of occurrence scores the relative probability that a given process step will cause a quality attribute to deviate outside of its acceptable limits. The occurrence score reflects how capable the process is to control a particular PQA within predetermined ranges. Information evaluated to determine the occurrence score includes the impact of the particular unit operation on the PQA, the level of PQA-specific process knowledge, sensitivity to raw material variability, variability within operating ranges as determined by univariate or multivariate (i.e., design space) process characterization, process monitoring capability analysis results (when available), and the level of redundancy to control the PQA in other unit operations. Scores are assigned for each unit operation, including intermediate hold steps.

A decision tree used to determine likelihood of occurrence scoring is presented in Figure 5. A letter code is used to document the path followed along the decision tree to arrive at a given score because it is possible to arrive at the same score through multiple logic pathways.

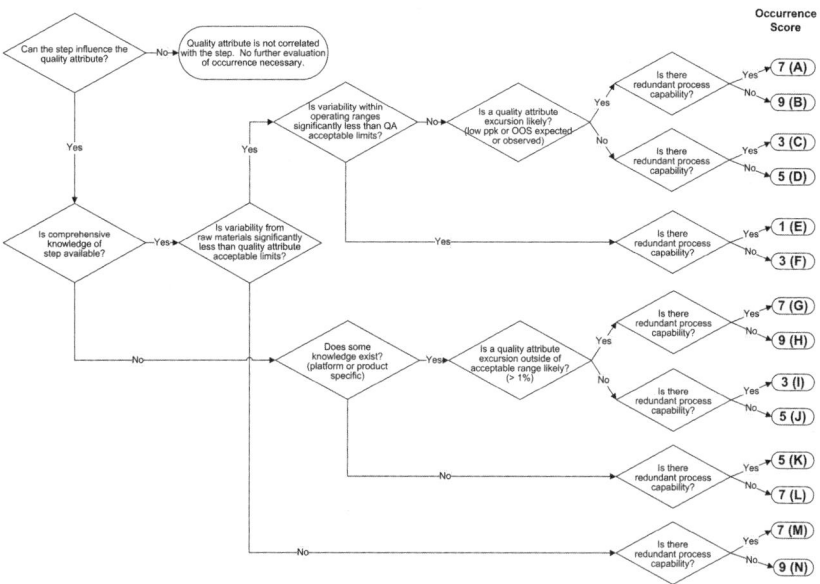

Figure 5. Likelihood of occurrence decision tree.

Early in the development lifecycle, comprehensive process knowledge is less likely. At this stage, scores may be based upon general platform understanding and/or limited process development data. A higher degree of uncertainty generally leads to higher scores, although the highest likelihood of occurrence score (here given as 9) is reserved for steps where excursions are expected or known to be likely and redundant process capability does not exist.

Detection (Methodology)

The ability to detect a PQA excursion and prevent its impact on the patient is reflected in the detection score. The detection score utilized in the PQRA presented here takes into account two dimensions of detection: analytical method capability and control stringency. Analytical method capability considers factors such as limit of quantitation, precision, specificity, and whether orthogonal methods would detect the same PQA. For example, a method measuring a general physical property with poor resolution (e.g., a charge-based separation with poorly resolved peaks) would score as a higher detection risk compared to a method designed to accurately resolve a specific product-related impurity.

The second dimension of detection considered in the assessment is the control stringency. Control stringency depends on the frequency of testing and what actions are driven based on the data obtained (i.e., the testing strategy). When a PQA is tested on each lot (e.g., as part of specification or in-process testing), there is a high likelihood that an excursion will be detected, which in turn results in a low detection score. Testing to support process characterization, validation, stability, and/or comparability is not performed routinely on every lot. Thus, testing performed to only support these activities is less likely to detect a PQA shift, indicating higher risk if no other testing is performed.

Tests with associated action limits provide assurance that deviations will be detected and investigated; therefore, such tests will result in lower scores than testing performed without predefined limits for the purpose of process understanding/evaluation. A test with an associated rejection limit represents an even stronger control because it represents a known product quality threshold that the company has committed to not exceed. In contrast, action limits ensure excursions will be investigated, but there is no firm commitment regarding the outcome of the investigation. Thus from a regulatory perspective, the detection risk can be considered lower for tests with associated rejection limits. The method capability and control stringency concepts are combined to establish a detection score. The detection score is determined from Table 4 by looking up a composite score associated with the method capability ("n") and control stringency ("i"). More detailed decision trees (not shown) can provide additional guidance as needed for determining the intermediate scores for method capability and control stringency.

Detection is scored at each unit operation as part of the PQRA exercise. However, it is recognized that in many instances detection may occur in a downstream processing step, and this downstream detection would be capable of detecting excursions due to upstream processing. In such instances, the downstream detection score may be propagated to upstream steps such that the overall detection score assigned to a unit operation accurately reflects the ability to detect a PQA excursion prior to patient impact. For example, most PQAs require partially purified samples for analysis; therefore, they are not detected at the production bioreactor step. However, excursions in product quality caused by the production bioreactor step would be detected through downstream in-process controls and/or specification testing.

The testing strategy is required as an input to determine the detection score and assess overall product quality risk. It should be noted that in practice, PQRA scoring can be used in an iterative fashion to refine the testing strategy such that an acceptable overall risk profile is ultimately achieved. For example, the frequency of testing may be increased to reduce the risk associated with an attribute initially identified as high risk. However, other mitigation actions could also be considered, such as improving process capability (to lower likelihood of occurrence score) or demonstrating low risk of patient impact (to lower the severity score). The risk assessment tool described in this chapter is designed to be used flexibly in an iterative fashion to refine the control strategy.

Table 4. Detection Scoring Matrix

	Detection Scoring	Control Stringency				
		Testing Every Lot with Reject Limits ($i = 1$)	Testing Every Lot with Action Limits ($i = 3$)	Testing Every Lot without Limits or Routine Periodic Testing with Limits ($i = 5$)	Periodic Testing ($i = 7$)	Characterization ($i = 9$)
Method Capability	Qualitative ($n = 9$)	5	6	7	8	9
	Low precision, quantitative ($n = 7$)	4	5	6	7	8
	Not orthogonal, nonspecific, precise ($n = 5$)	3	4	5	6	7
	Orthogonal, nonspecific, precise ($n = 3$)	2	3	4	5	6
	Specific, precise ($n = 1$)	1	2	3	4	5

Determination of Risk Level

The individual scores for severity, likelihood of occurrence, and detection can be combined to determine the overall risk level associated with the PQA. The first step in establishing overall risk is to utilize the severity and likelihood of occurrence scores to determine a preliminary hazard risk level (Table 5). The preliminary hazard level identifies process steps at greatest risk of impacting high-criticality PQAs. This information can be useful to prioritize process design (i.e., development and characterization) activities on the most impactful process steps. The preliminary hazard is also an important input for establishing a risk-based testing strategy. As described above, the testing strategy is used to establish the detection score.

Table 5. Preliminary Risk Hazard

Likelihood of Occurrence	*Severity of Risk*				
	Insignificant (1)	*Minor (3)*	*Moderate (5)*	*Major (7)*	*Severe (9)*
Frequent (9)	Medium	Medium	**High**	**High**	**High**
Likely (7)	Low	Medium	**High**	**High**	**High**
Occasional (5)	Low	Medium	Medium	**High**	**High**
Unlikely (3)	Low	Low	Medium	Medium	**High**
Remote (1)	Low	Low	Low	Low	Medium

The detection score and the preliminary hazard risk level are used to establish an overall risk level as shown in Table 6. Overall risk considers the ability to detect and control deviations to prevent impact to patients. Overall risk levels for patients can be low even for high preliminary hazard risk attributes if the ability to detect excursions and prevent their impact to patients is almost certain (e.g., a specific, precise analytical method is used as a release specification test for every lot).

Table 6. Overall Risk Hazard

Risk Level from Table 5	Detection								
	Almost Certain (1)	Very High (2)	High (3)	Moderately High (4)	Moderate (5)	Slight (6)	Remote (7)	Very Remote (8)	Absolutely Uncertain (9)
High	Low	Medium	Medium	Medium	**High**	**High**	**High**	**High**	**High**
Medium	Low	Low	Low	Low	Medium	Medium	**High**	**High**	**High**
Low	Low	Low	Low	Low	Low	Low	Medium	Medium	Medium

Lifecycle Application of the PQRA

The PQRA can be a useful tool to guide development of an effective, risk-based control strategy. Development of a control strategy is an iterative process of applying the PQRA at various stages of development. This enables the control strategy to be based on a fundamental understanding of how the process affects PQAs. This section describes the development of the control strategy at key points in the commercialization process.

Figure 6 illustrates key points in the commercialization process where the PQA Assessment and PQRA can be applied. Prior to initiating commercial process development, a PQA Assessment is performed to provide a list of all known PQAs and their associated severity scores. The output of the PQA Assessment is used by developers to ensure that process development is prioritized appropriately for robust control of high-criticality PQAs. Scoring an overall risk level using the PQRA requires the manufacturing process to be defined so that the likelihood of occurrence can be scored. Therefore, the PQRA is not performed until after the commercial process has been established. At the time of initial process lock, a PQRA is useful for identifying whether process improvements or redundancies may be necessary to further reduce risk. This assessment also serves to ensure the proper focus in process characterization by identifying the process steps most likely to impact PQAs. This assessment is useful for justifying the selection of potential critical performance parameters for evaluation during process characterization, and it can be used to ensure that analytical methods are developed for appropriate in-process pools. The starting point for the commercial process development assessment is a baseline (non-molecule specific) risk assessment performed for the platform process.

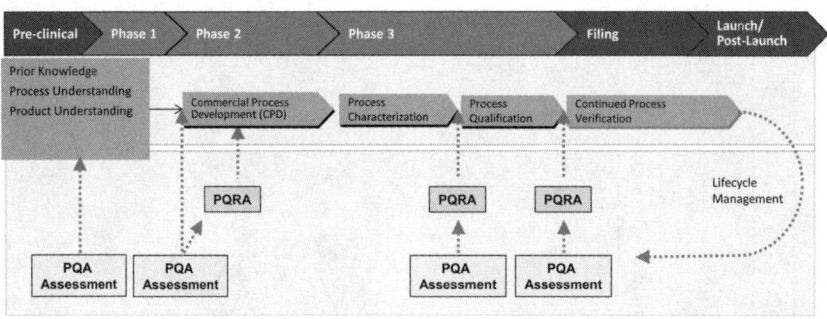

Figure 6. Risk assessments (Product Quality Risk Assessments [PQRAs] and Product Quality Attribute [PQA] Assessments) through the development lifecycle build to a final control strategy.

Following process characterization, the PQRA is repeated prior to initiating the process performance qualification (PPQ) phase of process validation. This is an important assessment that will be used to finalize the control strategy that will be in place for PPQ. Improved process understanding developed through process characterization will be applied to refine the procedural and testing controls (e.g., plans for comparability testing, in-process controls IPCs, validation). The assessment documents a prospective, risk-based approach to justify the PPQ test plan, including those parameters that will only be monitored during PPQ or comparability exercises (i.e., parameters to be "validated out").

After PPQ, the PQRA is repeated in preparation for the marketing application. This risk assessment supports the justification for specifications and control of critical steps and intermediates, and it is used to justify the test plan for routine manufacturing. The control strategy ultimately will be finalized based upon feedback from the file review with the appropriate regulatory agencies. Following approval, the risk assessment should be reevaluated on a periodic basis to confirm the appropriateness of the control strategy as more experience is gained with the process and product.

PQA Example 1: Fc HMGs

Introduction

Human serum IgG contains a conserved N-linked glycan on Asn 297 in constant domain 2 (C_H2) of the Fc region of each heavy chain (H). Recombinant therapeutic antibodies, in general, also possess glycans only at this site, as additional N-linked glycosylation sites are typically removed through DNA engineering. N-linked glycans are branched structures, and the types are defined by the monosaccharides at the end of the branches, as shown in Figure 7. In high mannose types, all branches terminate in mannose; in complex types, all branches terminate in sugars other than mannose (galactose, sialic acid, or *N*-acetylglucosamine); and in hybrid types, some branches terminate in high mannose, whereas others do not. Within each of these glycan types are a multitude of forms.

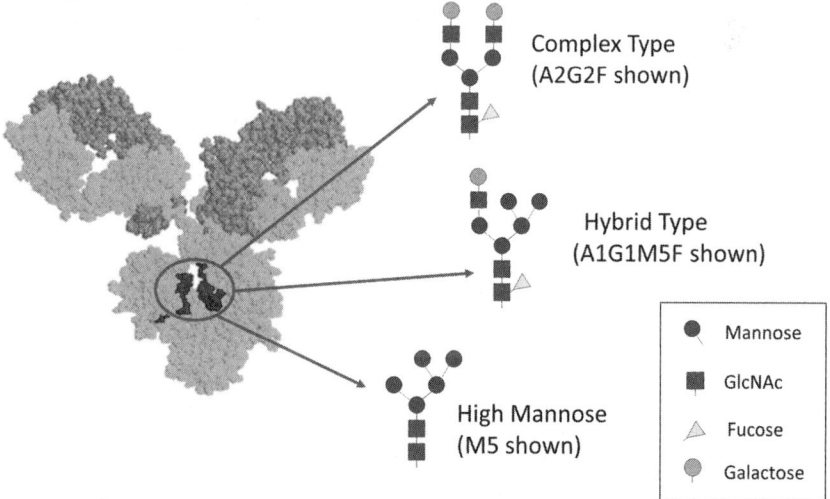

Figure 7. N-linked glycan types.

When selecting methods to analyze and characterize PQAs, consideration must be given to the existing knowledge of the glycobiology field and practical aspects and limitations during development. Many features of the Fc N-linked glycans have been found to impact safety and efficacy (4). Knowing these features and focusing on the analysis and control of relevant ones will improve the quality of the product and reduce cost of development. Fc glycans potentially could be considered as a single PQA, the relative levels of the three main types, or as a collection of more than 30 individual structures that comprise the three main types (5). By limiting analysis and control to the relative levels of the three main types, the developer would be ignoring features that affect safety or efficacy. Within the complex glycan type, the degree of fucosylation; the level of galactose-α-1,3-galactose linkages (also called α-Gal) (6); the type of sialylation; and the amount of terminal β-galactose all affect the product differently, and these effects may be product-specific. Therefore, a thorough review of how the Fc glycan features impact the specific product will improve the utility of the analytical approach and control strategy, and will reduce the risk to patients. In the case of Fc HMGs, the HMG structures 5 (Man5) through 9 (Man9) are considered to have similar impacts on safety and efficacy (structures shown in Figure 8), and therefore are considered as a collection of forms (7). Analytical methods would be developed that accurately measure the summed HMGs relative to the other Fc glycans. Discussion about glycan analytical methods and their specificity can be found in the Glycosylation chapter/Volume 2, Chapter 4.

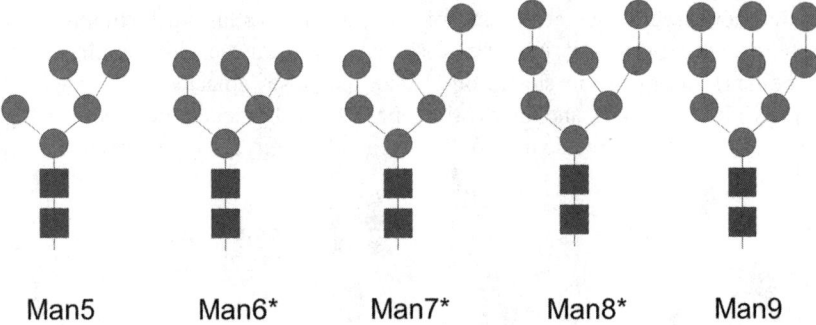

Figure 8. High-mannose structures. Asterisk denotes existence of multiple structural isoforms.

High-Mannose Attribute Impact

Safety

Attributes can affect the safety of the therapeutic antibody by increasing an undesirable activity of the drug, stimulating the drug's immunogenicity, or causing some other side effect. As is the case with many attributes, the safety of HMGs is inferred through indirect evidence, such as whether the attribute is naturally occurring, the degree of clinical exposure to the specific therapeutic antibody being studied or through experience with other therapeutic antibodies, and from animal studies. Supporting evidence can be obtained through cell-based immunogenicity predicting studies. The relative quality of the evidence should be considered when weighing the results from various studies.

The high mannose structures found on therapeutic antibodies (Figure 9) are naturally occurring. Low levels of Man5 (0.07%) have been measured on human serum IgGs obtained from healthy subjects (5). Both the glycan structure and the protein attachment site on the antibody are common to the endogenous and therapeutic antibody, indicating that this epitope would not be novel for a naïve patient. HMGs at levels of 2 to 15% have been found on commercial antibodies analyzed to date (not shown), but the levels do not correlate with the percentage of patients generating antidrug antibodies, suggesting that this attribute does not have a significant impact on the drug's overall immunogenicity rate.

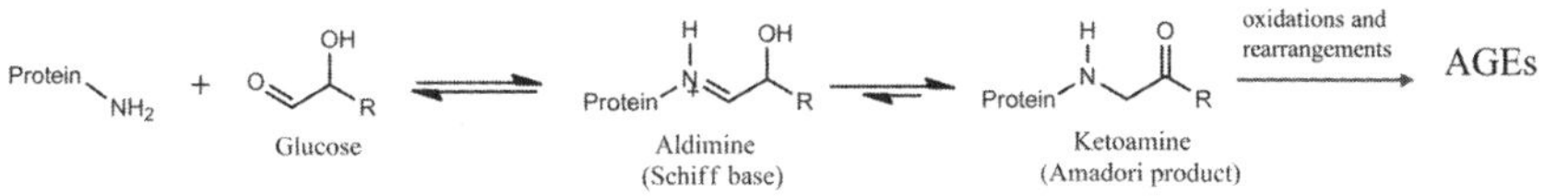

Figure 9. Glycation of proteins produces advanced glycation end products (AGEs).

Certain Fc glycan forms, particularly complex glycan types lacking core fucose, can significantly increase antibody-dependent cellular cytotoxicity (ADCC) of IgG1 therapeutic antibodies by improving binding to the Fc gamma receptor FcγRIIIa (8). If ADCC is not part of the MoA, enhanced binding of a therapeutic antibody to FcγRIIIa through attribute changes may lead to increases in off-target ADCC. Because HMGs lack core fucose, these Fc glycans result in higher ADCC activity than complex glycans containing core fucose (9). Therefore, depending on the core fucose levels, high mannose could potentially increase or decrease (if complex glycans are mainly afucosylated forms) off-target ADCC.

Efficacy

The efficacy of the therapeutic antibody *in vivo* is a function of its potency and dosage, which is affected by its PK. Because a therapeutic antibody's MoA may include target binding and effector functions, an attribute's effects on each of these activities should be considered. Fc glycan types have not been shown to affect the antibody epitope binding. As mentioned in the safety section, HMGs lack core fucose and, therefore, may increase the ADCC activity of a therapeutic antibody. Understanding the magnitude of these effects may require comparing antibodies enriched with different glycan forms using a cell-based ADCC assay developed specifically for the therapeutic antibody.

The effect of HMGs on monoclonal antibody (mAb) PK has been demonstrated experimentally by following compositional changes over time in humans (7, 10). Investigators found that the population of therapeutic antibodies with Man5 glycan decreases over time in patients with normal or low doses. This change is consistent with differential clearance of antibodies containing Man5 but not loss of Man5 through chemical conversion because the product of any conversions was not observed (7). One example of conversion can be seen with Man6 through Man9, the larger forms of HMGs. In the first couple days after dosing, rapid decreases in relative levels of these forms are observed, with a concomitant increase of Man5 (7, 11). These decreases in Man6 through Man9 and increases in Man5 could be replicated *in vitro* using human serum, confirming chemical conversions. This rapid conversion *in vivo* of larger high-mannose forms that are larger than Man5 supports considering the biological impact of high-mannose forms collectively.

Severity Scoring for mAbA High Mannose

To arrive at PQA Assessment severity scores, general attribute knowledge about the Fc HMG and product-specific attribute knowledge should both be considered. A hypothetical antibody will be used here to provide a practical example of how this information can be combined to arrive at reasonable scores. In this antibody example, mAbA is an IgG2 type that is produced in Chinese hamster ovary (CHO) cells and whose MoA is to bind a soluble ligand in the blood and prevent the ligand binding to a cellular receptor. Dosing is subcutaneous, the half-life in serum is 10 days, and the PK is considered relatively important to establish efficacy. High mannose forms exist within the Fc glycan collection of forms at 15% of the total. The PQA Assessment and the PQRA are being performed at the end of commercial development leading into a phase 3 clinical trial.

Fc HMGs commonly are found on commercial therapeutic antibodies, have been found on endogenous human antibodies (5), and were present on mAbA lots used in phase 1 and phase 2 clinical trials. No concerns have been raised about the immunogenicity of this glycan type. Non-immune safety concerns would also be considered low. The IgG2 isotype has lower ADCC activity than IgG1. Moreover, the antibody target is a soluble ligand and, therefore, would be unlikely to elicit an ADCC response.

PK characterization studies with mAbA found that the drug possessing the Man5 glycan appears to clear approximately two-fold faster than other forms. Calculations using these data show that mAbA containing HMGs clear ~1.5- to two-fold faster than the whole mAbA. Details as to how this calculation was performed are shown in Appendix A. The HMG form is not expected to affect target binding, but that was not tested specifically for mAbA.

An example of PQA Assessment scoring for mAbA high mannose PQA is provided in Table 7. The overall severity score for mAbA high mannose is 7, which is taken from the highest of the subscores. This value does not reflect the overall level of the attribute or the ability of the process to control the levels. The value instead could be considered the impact if the attribute levels were not controlled. Imagine the impact on mAbA clearance if HMG levels fluctuated between 1% and 50% lot to lot. Overall clearance, defined by the area under the curve (AuC), would differ by about 25%. This change would be considered a significant impact on mAbA efficacy. HMG levels may not have varied by this degree, but the actual levels and process variation would be considered later in the PQRA.

Table 7. Potential Product Quality Attribute (PQA) Assessment Scoring for mAbA High-Mannose Glycans (HMGs)

Attribute	Safety		Efficacy		Final (Highest) Score	Rationale for Severity Score
	Immunogenicity	Non-Immune Safety	PK	Potency		
Fc HMGs	1	1	7	1	7	Impact to serum clearance

Likelihood of Occurrence

HMGs are intermediates in the synthesis of complex-type glycans. Hence, high mannose on recombinant mAbs is a result of incomplete Fc glycan processing in CHO cells. Few changes typically are observed in the uncharged glycan forms through purification processes as these steps were not designed for such purposes. However, enriched levels of HMGs can sometimes be observed associated with high molecular weight aggregate species (*12*). Presumably, intracellular aggregation can lead to incomplete glycan processing because high mannose itself does not appear to promote extracellular aggregation for IgG1 or IgG2 antibodies (*13*). In such instances, clearance of aggregates during purification may lead to measurable changes in the relative amount of HMGs.

Changes in HMG levels are more typically correlated with cell culture steps. In some cases, high mannose levels may be monitored as part of the clone selection process, because variability in relative HMG levels often is observed across clones. Cell culture conditions in the production bioreactor such as osmolality, culture duration, and trace element concentrations (e.g., Mn) also can affect HMG levels (*14*).

For our hypothetical case study molecule, we will assume that mAbA has relatively low levels of high molecular weight aggregates, and the proportion of HMG species remains relatively unchanged from the protein A pool through to drug substance. HMG levels were monitored as part of final clone selection to ensure the final clone selected generated an average of ≤ 20% HMGs. High mannose also was monitored during development of the commercial process, and relatively minor changes were observed when parameters such as pH, temperature, media composition, and feed strategy were optimized. Cell culture media are chemically defined, and no impact was noted from lot-to-lot variations in media components.

The first step in scoring likelihood of occurrence is to identify the steps that are correlated with the attribute. For mAbA, the clone selection and production bioreactor steps are the only unit operations directly correlated to HMG levels and, therefore, are the only steps that will be scored for occurrence in the PQRA. The likelihood of occurrence score next is determined for each step using the flowchart in Figure 5. For the clone selection step, comprehensive knowledge is available because all final clones were evaluated for HMGs. No variation was noted due to raw materials, and the final clone was selected specifically to ensure acceptable levels of HMGs. Because HMG levels do not change through purification, there is no downstream process capability to modify HMG levels. Therefore, according to Figure 5, an occurrence score of "3" is assigned to the step. The code "F" is recorded to indicate the path taken in the decision tree to arrive at this score.

For the production bioreactor, some process knowledge exists from the commercial process development experimentation. At this stage of development, however, the process has not yet been fully characterized using multivariate design of experiments (DOE), and; therefore, the knowledge is not comprehensive. The available data suggest that excursions outside of acceptable levels are not likely. Hence a score of "5" is assigned per Figure 5, and the code "J" is recorded to document the decision tree path.

Detection

Several analytical strategies are available to monitor high mannose levels. These include glycan mapping of the released glycan, peptide mapping with MS analysis of the glycopeptide, and assays specifically designed to monitor high-mannose forms. Glycan mapping is described in the Glycosylation chapter/Volume 2, Chapter 4. Various separation techniques can be employed for the labeled glycans. Peak characterization with MS is often used to ensure that peaks containing high mannose are not confounded with other glycan forms. Mass spectrometric detection is required when using peptide mapping to monitor high mannose because reversed phase (RP) chromatography is usually insufficient to resolve the numerous Fc glycan species. Sufficient mass resolution and MS/MS characterization of the resolved masses is required to ensure high-mannose specificity. Finally, alternative HMG monitoring techniques could include analytical Conconavalin A affinity chromatography or an assay combining a high mannose-specific endoglycosidase with a separation technique (15).

For the mAbA case study example, we will assume that glycan mapping is performed, and the peaks containing HMGs are well resolved. Because the glycan map analysis requires a partially purified sample for analysis, it is typically performed on the protein A pool for characterization or on the drug substance as part of comparability testing.

Table 4 is used to determine the detection score for the relevant process steps. Because detection requires a partially purified sample, there is no detection at the clone screening or production bioreactor steps. Detection at these steps is scored as "9" to reflect the lack of detectability. However, downstream testing is performed as part of drug substance comparability testing. The test method used at drug substance is specific and precise; therefore, method capability is scored as n = 1 per Table 4. Because testing is performed periodically as part of comparability, control stringency is scored as i = 7. Per Table 4, the overall detection score at drug substance is, therefore, 4. Because downstream detection at drug substance is capable of detecting changes in HMG levels resulting from the cell culture steps, the detection score of 4 is propagated as the overall detection score for the upstream steps.

Determination of Risk Level

Severity, likelihood of occurrence, and detection scores are used to determine the overall risk level. First, the severity and likelihood of occurrence scores are used to determine the preliminary hazard level per Table 5. The detection score is then combined with the preliminary hazard level per Table 6 to determine the overall unit operation risk. The quality attribute overall risk level is defined as the highest of the individual unit operation risk levels. A summary of the example scoring for our mAbA high mannose is shown below in Table 8. In this example, the high preliminary hazard risk level at the production bioreactor is mitigated by comparability testing with a specific, precise assay. The risk could be further mitigated by reducing the likelihood of occurrence score (e.g., by gaining additional process understanding through process characterization) or through enhanced detection (e.g., by adding a routine in-process control or specification test).

Table 8. Product Quality Risk Assessment (PQRA) Scoring for mAbA High Mannose [a]

Unit Operation	Correlation (↓, ↑ or Testing Only)	PQA Severity	Likelihood of Occurrence		Preliminary Hazard Risk Level	Detection at Unit Operation				Downstream Detection		Overall Unit Operation Risk Level
			Decision Tree Code	Likelihood of Occurrence Score		Detection Method	Capability (n)	Stringency (i)	Detection Score	Detected Downstream (if yes, list step)?	Overall Detection Score	
Clone/cell line	↓↑	7	F	3	Medium	Not detected at this step	9	9	9	Yes, detected at DS	4	Low
Production bioreactor	↓↑	7	J	5	High	Not detected at this step	9	9	9	Yes, detected at DS	4	Medium
Drug substance (DS)	Testing only	7				Glycan map, comparability test	1	7	4			

[a] Quality attribute overall risk level is medium.

Product Quality Attribute Example 2: Glycation

Introduction

Glycation is a non-enzymatic protein glycosylation with a reducing sugar. The reaction occurs between primary amines on the protein, either through lysine side chains or the N-terminal amine and the carbonyl carbon of the sugar. The initial product is a Schiff base that can rearrange to form a more stable ketoamine (*16, 17*), as illustrated in Figure 9. Oxidation and further rearrangement of the ketoamine and aldimine products can generate a collection of stable adducts known as advanced glycation end products (AGEs). Because these reactions are slow, naturally occurring AGEs typically are observed on long-lived proteins such as those in the eye lens (*17*).

Recombinant therapeutic antibodies display glycation at relatively low levels of 0.1 to 0.3 mol sugar/mol antibody. Most analyses list the amount as total glycation, which is the amount found on the protein, and not the amount on each individual lysine site on the protein. These species are either aldimines or ketoamines that form as the result of glucose addition. Glycation on therapeutic proteins occurs when the protein is secreted from cells into media containing glucose, the primary sugar feed. No AGE modifications have as yet been described on therapeutic antibodies, likely due to the limited time that they interact with glucose in production reactors.

The glycation sites on the protein primary sequence have been analyzed for a number of proteins. On some proteins, specific lysines are modified to a much greater degree, indicating sequence or structural specificity to the reaction (*18, 19*). In general, the amine pK_a and lysine solution exposure determine the degree of Schiff base formation, and the proximity of amino acids capable of proton abstraction promote ketoamine formation (*18, 20*). Buffer components that can bind near the reaction site and abstract protons also can affect stable product formation rates (*21*). Depending on relative site reactivities for a therapeutic antibody, glycation could be found primarily at one site or well distributed over the many lysine positions in the molecule.

Safety

As described in the HMG example, attributes can affect the safety of the therapeutic antibody in multiple ways. As for other attributes, much of the evidence concerning the impact of glycation is indirect, involving knowledge about the existence of the attribute in clinical trials and on endogenous antibodies. Because AGEs have not been detected on therapeutic antibodies to date, their impact on safety and efficacy is not evaluated in this glycation section.

Upon injection of therapeutic IgGs into humans, relative glycation levels were found to increase as a function of time (*22*). Changes to the light chain (L) were low but easily detectable at 0.00092 glucose additions per chain per day. Levels of glycation found on the Fc portion of endogenous IgG isolated from healthy subjects were on average 0.045 glucose molecules per fragment. From these two studies, the *in vivo* glycation rates on antibodies was estimated to be 0.006 mol glucose per mol antibody per day, or about 0.14 mol of glucose per mol of endogenous antibody with a 23-day circulating half-life. Other published studies have measured glycation levels on human antibodies from healthy individuals (*23–25*). However, results from these studies vary widely. Normal antibody glycation values from the Kaneshige study (*25*), which used two non-MS-based techniques, can be converted to an average glycation rate of 0.14 Glc/antibody (i.e., about 14% of the antibody molecules have a single glycation), which is in close agreement with the study described above.

In principle, glycation could affect non-immune safety through increasing off-target binding or increasing non-MoA ADCC. However, no such changes in activity have been reported.

Efficacy

Antibody binding to its target has the potential to be affected by glycation, but the effect will depend on the protein sequence in and around the variable complementarity-determining region (CDR), which varies among antibody molecules. The impact on potency can be determined through *in vitro* forced glycation, which has the potential to generate much higher levels of site-specific glycation than can be achieved during production.

For an antibody with a global glycation level of 14% (0.14 mol glycation/mol antibody), approximately 14% of the antibodies have a single attached glucose. Because there are several reactive lysines on the Fc and there are reactive lysines throughout the rest of the molecule, the glycation level at any specific lysine involved in an Fc function would be a fraction of the overall glycation level. Through forced glycation conditions, however, all reactive lysine sites on therapeutic antibodies were glycated to an appreciable degree. In this study, IgG1 antibodies containing on average > 40 glucose molecules per antibody were unaffected in binding to FcγRIIIa, suggesting ADCC activity would not be altered.

To determine the impact of glycation on IgG PK, highly glycated IgG1 and IgG2 were prepared containing on average 42 to 49 glucose molecules per IgG. The neonatal Fc receptor FcRn increases the circulating half-life of antibodies through a salvage pathway, so its binding at low pH is used as a surrogate for attribute effects on PK. Binding to FcRn was similar or identical to the non-glycated IgG controls. Although the modifications were well distributed throughout the protein sequence, no changes in the tested Fc functions were observed (*22*).

Severity Score for mAbA Glycation

General attribute knowledge about glycation, as well as product-specific attribute knowledge, should be considered when arriving at severity scores. The same hypothetical antibody used in the Fc high mannose attribute example, mAbA, will be used here to arrive at product-specific scores for glycation severity. All the antibody features still apply (i.e., mAbA is an IgG2 type produced in

CHO cells whose MoA is to bind a soluble ligand in the blood and prevent it binding to a cellular receptor). Dosing is subcutaneous, and the PK half-life in serum is 10 days at the relevant dosing. To this is added information relevant to the glycation assessment. The overall glycation levels for the drug substance is 10 to 15% overall, meaning that mAbA has 0.1–0.15 mol of glycation per mol of antibody. Characterization shows that the glycation is distributed over several lysine sites in the molecule.

Glycated mAbA was present in the lots used in clinical trials to date. Furthermore, additional glycation of mAbA is expected to occur after injection into patients at rates similar to endogenous antibodies based on comparing *in vitro* glycation of mAbA to test antibodies that had been studied *in vivo*. Therefore, patients have previously been exposed to mAbA glycation epitopes. No published studies have linked glycation with immunogenicity for therapeutic or endogenous antibodies. Overall, glycation is not expected to impact mAbA's safety.

Glycation levels on mAbA would never be expected to approach those necessary to affect target binding. With global glycation levels at 0.15 mol glycation/mol antibody and equal glycation at 10 lysine sites, the glycation at any specific site would only be 0.015 mol glycation/mol antibody. *In vitro* forced glycation experiments with mAbA show that extremely elevated levels of 1.5 mol glycation/mol antibody did not affect the *in vitro* potency assay.

No direct experiments were performed to test the impact of glycation on mAbA clearance in humans. However, extremely high levels of glycation do not impact FcRn binding for multiple IgG1 and IgG2 test molecules (7). Because these binding sites are conserved between IgG2 molecules, the results can be applied to mAbA. Without effects on target or FcRn binding, no effects on PK are expected.

A potential PQA Assessment scoring table for mAbA glycation is shown in Table 9. The overall score is glycation on mAbA is 1, indicating no or very low safety and efficacy impact. Again, this value does not reflect the overall level of the attribute or the processes ability to control the levels. With this low score, a PQRA may not be performed for this attribute, as loss of control or poor detection may not increase risk significantly. To demonstrate how changes process control or detection would impact the risk of this low-scoring attribute, PQRA scoring changes are described below.

Table 9. Product Quality Attribute (PQA) Assessment Scoring for mAbA Glycation

Attribute	Immuno-genicity	Non-immune Safety	PK	Potency	Final (Highest) Score	Rationale for Severity Score
Glycation	1	1	1	1	1	Naturally occurring attribute with no impact on potency

Likelihood of Occurrence

Antibody glycation is generated spontaneously in the cell culture media in the presence of glucose. Cell culture conditions, such as pH, glucose concentration, and the time the antibody spends in the culture, all affect glycation levels on the antibody. Amines in culture will compete with the protein for the Schiff base formation, so variations in amino acids may also change glycation levels. Because glycation occurs through a chemical reaction in the culture supernatant, it can also occur in the absence of cells, for example, during harvest processing and during harvest pool hold steps. Formation of a stable lysine-sugar glycated product neutralizes the positive charge of the lysine amine, shifting the protein to a more acidic form. Thus, charge-based separation techniques have the potential to reduce glycation levels on the product, although this is not often observed in practice for industrial-scale ion exchange chromatography. With the low levels described here, a high percentage of the glycated forms would result is a negative charge over the unglycated species.

For our mAbA case study molecule, glycation is known to occur in the production bioreactor. Furthermore, changes observed through CEX during harvest pool hold studies were shown to be due to increases in glycation. No changes were observed in glycation through downstream chromatography steps. Therefore, the process steps correlated with glycation include the production bioreactor, harvest, and harvest pool hold steps.

Applying the likelihood of occurrence decision tree in Figure 5, the production bioreactor is scored as "5 (J)." Although charge variants were monitored during commercial process development and found to be relatively consistent, the process has not yet been fully characterized and changes in net charge can be confounded with other modifications such as deamidation. However, the data are sufficient to conclude that excursions beyond acceptable ranges are unlikely. Changes in charge variants observed during harvest development resulted in additional targeted process characterization. The data indicate that significant changes in glycation are possible if operating conditions such as pH and hold times are not well controlled. As a result of the additional characterization, procedural controls were put in place to effectively limit potential glycation. In this case, comprehensive knowledge is available, and the likelihood of occurrence is scored as "3 (F)." The procedural controls are sufficient to ensure low levels of glycation in the harvest steps.

Detection

Two general strategies are employed to monitor protein glycation: global glycation analysis and site-specific glycation analysis. Site-specific glycation analysis will determine the level at specific lysine positions on the primary sequence, usually through peptide mapping with MS. Proteases that do not cleave at lysines are often used for this purpose, as sugar attachment can inhibit trypsin or Lys-C cleavage. This analytical approach is best used when the glycation occurs to a large degree at a key site, such as on a lysine in the CDR. In global glycation analyses, site-specific information cannot be obtained, but the method is often more sensitive because the low-level glycation at many sites are summed. These analyses include RP high-performance liquid chromatography (RP-HPLC) with or without MS, and boronate affinity chromatography. Prior removal of the Fc glycan by endoglycosidase digestion improves the accuracy of these methods. Changes in glycation can also be detected indirectly through methods such as cation exchange HPLC (CEX-HPLC) that detect changes in net charge. Glycated species are present in the acidic peaks, often confounded with other modifications such as deamidation.

For our case study example, mAbA glycation levels were monitored in characterization studies by whole mass RP-HPLC/MS analysis for global glycation levels. For routine monitoring, CEX-HPLC was used to monitor charge variants, although no action or rejection limits were applied. Both of the detection methods can be scored using Table 4 to determine whether one of the methods results in a more robust control strategy. The RP-HPLC/MS method is specific and precise ($n = 1$), but it is used for characterization only ($i = 9$), and therefore the detection score is 5. In contrast, the CEX-HPLC method is nonspecific because glycation is not resolved from other modifications such as deamidation ($n = 5$), but it is routinely applied ($i = 5$). Hence, the overall detection for this method also scores as 5. Based upon the PQRA scoring, a specific method applied for characterization scores the same as a nonspecific method applied routinely. The detection scores could be reduced if desired by applying limits to the routine testing or by applying the more specific method more frequently, for example, as part of comparability testing.

Determination of Risk Level

A summary of the scoring for mAbA glycation is shown below in Table 10. In this example, the information available is sufficient to conclude low preliminary hazard for all process steps. The low risk reflects the low attribute criticality (severity score of 1) and sufficient process understanding to conclude a low risk of excursions. For such an attribute, a characterization-only control strategy is sufficient to ensure low risk to patients.

Table 10. Product Quality Risk Assessment (PQRA) Scoring for mAbA Glycation [a]

Unit Operation	Correlation (↑, ↓, or Testing Only)	Product Quality Attribute (PQA) Severity	Likelihood of Occurrence		Preliminary Hazard Risk Level	Detection at Unit Operation				Downstream Detection			Overall Unit Operation Risk Level
			Decision Tree Code	Likelihood of Occurrence Score		Detection Method	Capability (n)	Stringency (i)	Detection Score	Detected Downstream (if yes, list step)?	Overall Detection Score	
Production bioreactor	↑	1	J	5	Low	Not detected at this step	9	9	9	Yes, detected at DS	5	Low
Harvest	↑	1	F	3	Low	Not detected at this step	9	9	9	Yes, detected at DS	5	Low
Harvest pool hold	↑	1	F	3	Low	Not detected at this step	9	9	9	Yes, detected at DS	5	Low
Drug substance (DS)	Testing only	1				Reversed phase high-performance liquid chromatography/mass spectrometry (RP-HPLC/MS) for characterization only	1	9	5			

[a] Quality attribute overall risk level is low.

This example highlights the value of enhanced product understanding. Historical control strategies may have relied upon release specifications for charge-variant minor species, given the relative lack of knowledge regarding the criticality of the attribute(s) being monitored. Improved understanding of the relative criticality of the molecular attributes can be combined with process understanding to develop more efficient control strategies.

Conclusion

This chapter provides an example of how to conduct an appropriate risk assessment for a process control strategy using QbD principles. Key features for any approach would include focusing on high-criticality quality attributes and the process steps affecting these, as well as an understanding of the assay performance. Because QbD requires a thorough understanding of product and quality attributes, proper assays should be applied appropriately to control quality attributes to ensure consistent quality. Successful application of PQRA requires engagement and participation from a cross-disciplinary team at key stages of the product lifecycle. As product and process knowledge matures through the product lifecycle, the PQRA provides a solid framework to document understanding, evaluate new information, and adjust the control strategy accordingly.

Benefits of applying the PQA Assessment and PQRA methodology include compliance with regulatory expectations, more efficient process development, and streamlined control strategies. ICH Q10 establishes the expectation that quality risk management is used to establish control strategies (*26*). The PQA Assessment and PQRA approach can be used to comply with this expectation, and they also can be a useful framework for describing and justifying the control strategy in the marketing application. Applying the PQRA at key stages of commercialization helps to focus development on the areas of greatest risk, which improves efficiency. Streamlined control strategies ultimately contribute to lower cost of goods by eliminating redundant, non-value-added testing.

Acknowledgments

We would like to thank a number of people who have contributed to the development of the assessment tools describe here. In particular, we acknowledge the efforts of Drew Kelner, Paul Tsang, Darrin Cowley, Mike Moxness, Brent Kendrick, Yijia Jiang, Erwin Freund, Carol Krantz, Paul Konefal, Paola Schick, Duane Bonam, Mark Benke, David Hambly, Melissa Weakly, Tom Monica, Rob Woolfenden, Cenk Undey, Paul Konold, Karen Schlobohm, Kathy Leach, Nolan Polson, Rebecca Spohn, Chulani Karunatilake, Tamas Blandl, Claudia Baikalov, Carole Heath and Bob Kuhn. We also thank Tony Mire-Sluis, Barry Cherney, Joseph Phillips, Aine Hanly, and Duane Bonam for critically reading the chapter.

Appendix A: Estimating the Impact of High-Mannose Glycans on Drug Clearance

Antibodies with HMGs typically clear faster than ones with complex glycans (*7, 10*). Preparations of the antibody drug may contain a mixture of these glycans; therefore, knowing how the glycan form levels affect clearance is necessary to estimate targets for process development and for specification setting. One way to estimate the clearance impact is through the use of glycan analysis of the drug taken from serum samples in human PK studies. If the antibody containing HMGs clears faster, the glycan fraction with high mannose will be seen to decrease over time (see Figure A-1). Using this information, the effect with different theoretical levels of high mannose can be calculated (*3*).

In the example shown below on the left, the overall PK profile is given for a subcutaneous injection of mAbA in humans.

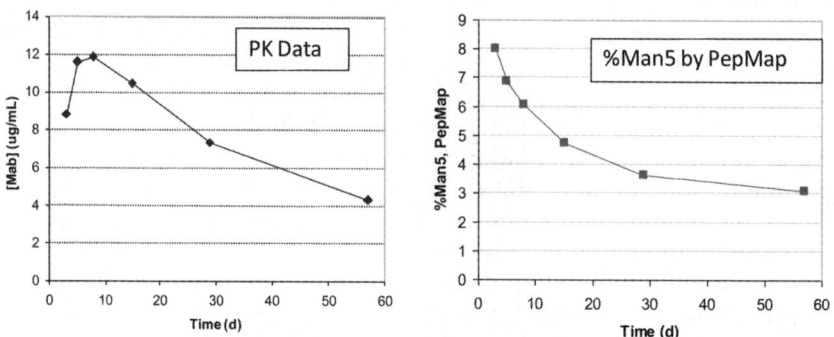

Figure A-1. mAbA pharmacokinetic (PK) plot (left) and in vivo changes in relative high mannose glycans (Man5) in mAbA (right).

At each PK time point, the antibody is isolated from the serum sample and analyzed for glycan content. Various glycan analysis methodology can be used for this purpose. Peptide mapping and fluorescent labeling with HPLC analysis of released glycans are two common methods with the requisite sensitivity and selectivity for this type of study. The relative glycan composition over time *in vivo* can then be generated (Figure A-1, right). The composition of most glycans is stable, whereas HMGs such as the Man5 glycan shown in the figure decrease over time. Other glycans as a group would increase, because of the following relationship:

$$\text{fraction}_{HMGs} + \text{fraction}_{other\ glycans} = 1$$

With the PK data and the glycan composition data, the individual PK profiles can be constructed for the drug containing high mannose and the drug containing other glycans ($PK_{total} - PK_{HMG}$; see Figure A-2).

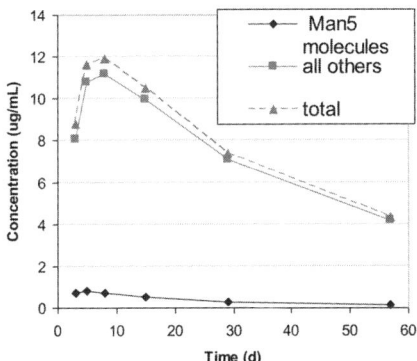

Figure A-2. Calculated mAbA pharmacokinetic (PK) curves with high mannose (Man5 molecules) and other glycans (all others). The combined PK curve is the same as the total antibody PK curve.

The AuC is determined for each of these plots: AuC_{HMG} and AuC_{other}. Together these sum to AuC_{total}. The relative impact of these glycan forms on clearance then are obtained by dividing the AuC by the relative glycan level (e.g., AuC_{HMG}/Fraction HMG), which here is called AuC_{HMG}/dose. Finally, the difference in overall clearance can be calculated by substituting different glycan compositions in the following calculation:

$$AuC_{total} = Fraction_{HMG} \times (AuC_{HMG}/dose) + Fraction_{other} \times (AuC_{other}/dose)$$

In this way, differences in PK can be calculated for two different theoretical levels of high mannose. Judgments then can be made as to the proper limits on the HMG ranges for lot release.

Appendix B: Estimating Patient Exposure to PQAs Converting *in Vivo*

Several PQAs shown to change over time during cell culture and in drug product storage have also been found to convert in the body. *In vivo* studies, previously described and outlined in Figure 4, can be used to estimate the exposure of the patient to the attribute (2, 27). The overall exposure is dependent on the initial levels, the conversion rate, and the drug clearance. At each PK time point, the antibody is isolated from the serum sample and analyzed for the attribute. Both forms of the attribute—reactant and product—are determined, so that the relative levels of these can be calculated. In the example shown here, mAbB (IgG1) was injected intravenously into humans and studied for thioether conversion. Overall drug levels *in vivo* were determined using an ELISA assay and are shown in Figure B-1 on the left.

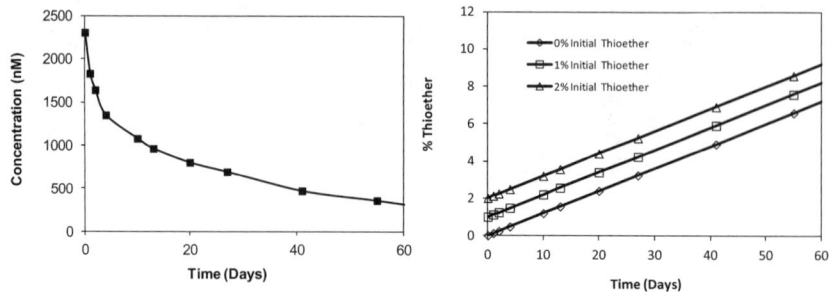

Figure B-1. mAbB pharmacokinetic (PK) plot (left) and relative thioether levels in vivo (right).

The affinity-purified mAbB was digested with Lys-C under non-reducing conditions, and the disulfide-linked peptides were analyzed in a peptide map (*28*). One peptide, the peptide containing the L-H disulfide bond, is prone to convert to a thioether bond (*29*). Relative levels at this site are then calculated at a fraction of total L-H peptide (Figure B-1, right). Relative thioether levels increased linearly with time *in vivo*. The impact of changes in the initial thioether levels on the relative levels *in vivo* can be calculated and compared, as shown. In the thioether case, the changes are additive and result in time courses with parallel lines.

Patient exposure to the attribute can then be calculated for each case of the drug differing in the initial level of thioether. The PK drug curve (Figure B-1, left) is multiplied by the relative thioether level plots (Figure B-1, right) for each initial thioether level scenario to generate a series of theoretical patient exposure curves, shown in Figure B-2. Overall patient exposure for each case can be expressed as the AuC. For this example, an increase of thioether level in the drug from 1% to 2% (a 100% increase in the PQA level in the drug) results in an increase in patient exposure of approximately 21%. Generating a large amount of mAbB with 1% thioether as compared to one with no thioether only results in an increased patient exposure of 25%.

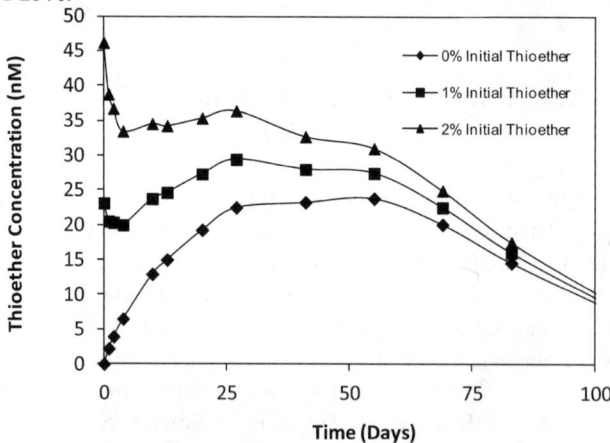

Figure B-2. Calculated thioether concentrations from mAbB in vivo with different initial relative thioether levels.

References

1. ICH Harmonised Tripartite Guideline. Pharmaceutical Development: Q8(R2), 2009. http://www.ich.org/fileadmin/Public_Web_Site/ICH_Products/Guidelines/Quality/Q8_R1/Step4/Q8_R2_Guideline.pdf.
2. Liu, Y. D.; van Enk, J. Z.; Flynn, G. C. *Biologicals* **2009**, *37*, 313–322.
3. Goetze, A. M.; Schenauer, M. R.; Flynn, G. C. *mAbs* **2010**, *2*, 1–8.
4. Jefferis, R. *Biotechnol. Prog.* **2005**, *21*, 11–16.
5. Flynn, G. C.; Chen, X.; Liu, Y. D.; Shah, B.; Zhang, Z. *Mol. Immunol.* **2010**, *47*, 2074–2082.
6. Chung, C. H.; Mirakhur, B.; Chan, E.; Le, Q. T.; Berlin, J.; Morse, M.; Murphy, B. A.; Satinover, S. M.; Hosen, J.; Mauro, D.; Slebos, R. J.; Zhou, Q.; Gold, D.; Hatley, T.; Hicklin, D. J.; Platts-Mills, T. A. *N. Engl. J. Med.* **2008**, *358*, 1109–1117.
7. Goetze, A. M.; Liu, Y. D.; Zhang, Z.; Shah, B.; Lee, E.; Bondarenko, P. V.; Flynn, G. C. *Glycobiology* **2011**, *21*, 949–959.
8. Shields, R. L.; Lai, J.; Keck, R.; O'Connell, L. Y.; Hong, K.; Meng, Y. G.; Weikert, S. H.; Presta, L. G. *J. Biol. Chem.* **2002**, *277*, 26733–26740.
9. Kanda, Y.; Yamada, T.; Mori, K.; Okazaki, A.; Inoue, M.; Kitajima-Miyama, K.; Kuni-Kamochi, T.; Nakano, R.; Yano, K.; Kakita, S.; Shitara, K.; Satoh, M. *Glycobiology* **2007**, *17*, 104–118.
10. Alessandri, L.; Ouellette, D.; Acquah, A.; Rieser, M.; Leblond, D.; Saltarelli, M.; Radziejewski, C.; Fujimori, T.; Correia, I. *mAbs* **2012**, *4*, 509–520.
11. Chen, X.; Liu, Y. D.; Flynn, G. C. *Glycobiology* **2009**, *19*, 240–249.
12. Strand, J.; Huang, C. T.; Xu, J. *J. Pharm. Sci.* **2013**, *102*, 441–453.
13. Lu, Y.; Westland, K.; Ma, Y. H.; Gadgil, H. *J. Pharm. Sci.* **2012**, *101*, 4107–4117.
14. Pacis, E.; Yu, M.; Autsen, J.; Bayer, R.; Li, F. *Biotechnol. Bioeng.* **2011**, *108*, 2348–2358.
15. Chen, X.; Tang, K.; Lee, M.; Flynn, G. C. *Electrophoresis* **2008**, *29*, 4993–5002.
16. Bunn, H. F.; Higgins, P. J. *Science* **1981**, *213*, 222–224.
17. Bucola, R.; Cerami, A. *Adv. Pharmacol.* **1992**, *23*, 1–33.
18. Zhang, B.; Yang, Y.; Yuk, I.; Pai, R.; McKay, P.; Eigenbrot, C.; Dennis, M.; Katta, V.; Francissen, K. C. *Anal. Chem.* **2008**, *80*, 2379–2390.
19. Miller, A. K.; Hambly, D. M.; Kerwin, B. A.; Treuheit, M. J.; Gadgil, H. S. *J Pharm. Sci.* **2011**, *100*, 2543–2550.
20. Venkatraman, J.; Aggarwal, K.; Balaram, P. *Chem. Biol.* **2001**, *8*, 611–625.
21. Watkins, N. G.; Neglia-Fisher, C. I.; Dyer, D. G.; Thorpe, S. R.; Baynes, J. W. *J. Biol. Chem.* **1987**, *262*, 7207–7212.
22. Goetze, A. M.; Liu, Y. D.; Arroll, T.; Chu, L.; Flynn, G. C. *Glycobiology* **2012**, *22*, 221–234.
23. Austin, G. E.; Mullins, R. H.; Morin, L. G. *Clin. Chem.* **1987**, *33*, 2220–2224.

24. Lapolla, A.; Fedele, D.; Garbeglio, M.; Martano, L.; Tonani, R.; Seraglia, R.; Favretto, D.; Fedrigo, M. A.; Traldi, P. *J. Am. Soc. Mass Spectrom.* **2000**, *11*, 153–159.
25. Kaneshige, H. *Diabetes* **1987**, *36*, 822–828.
26. ICH Harmonised Tripartite Guideline. Pharmaceutical Development: Q10, 2008. http://www.ich.org/fileadmin/Public_Web_Site/ICH_Products/Guidelines/Quality/Q10/Step4/Q10_Guideline.pdf.
27. Liu, Y. D.; Goetze, A. M.; Bass, R. B.; Flynn, G. C. *J. Biol. Chem.* **2011**, *286*, 11211–11217.
28. Zhang, Q.; Schenauer, M. R.; McCarter, J. D.; Flynn, G. C. *J. Biol. Chem.* **2013**, *288*, 16371–16382.
29. Tous, G. I.; Wei, Z.; Feng, J.; Bilbulian, S.; Bowen, S.; Smith, J.; Strouse, R.; McGeehan, P.; Casas-Finet, J.; Schenerman, M. A. *Anal. Chem.* **2005**, *77*, 2675–2682.

Appendix

Table 1. Acronyms

Acronym	Definition
2-AA	anthranilic acid
2-AB	2-aminobenzamide
AARS	aminoacyl-tRNA synthetases
ADC	antibody drug conjugate
ADCC	antibody-dependent cellular cytotoxicity
AEX	anion exchange chromatography
AF4	asymmetric flow field-flow fractionation
AFM	atomic force microscopy
AGE	advanced glycation end product
APC	antigen-presenting cell
API	active pharmaceutical ingredient
ASGPR	asialoglycoprotein receptor
Asn	asparagine
Asp	aspartic acid
AUC	analytical ultracentrifugation
AuC	area under the curve
CAGR	compound annual growth rate
CD	circular dichroism
CDC	complement-dependent cytotoxicity
cDNA	complementary DNA
CDR	complementarity-determining region
CE	capillary electrophoresis
CEX	cation exchange chromatography
CFR	Code of Federal Regulations
cGMP	current good manufacturing practice

Continued on next page.

© 2014 American Chemical Society

Table 1. (Continued). Acronyms

Acronym	Definition
C_H1, C_H2, C_H3	constant domains 1, 2 and 3 of antibody heavy chain
CHO	Chinese hamster ovary
cIEF	capillary isoelectric focusing
C_L	constant domain of antibody light chain
CMC	chemistry, manufacturing, and controls
CMP	cytidine monophosphate
CQA	critical quality attribute
CRM	certified reference material
cSDS	capillary sodium dodecylsulfate electrophoresis
CTD	Common Technical Document
Cys	cysteine
CZE	capillary zone electrophoresis
d	denatured
DHFR	dihydrofolate reductase
DLS	dynamic light scattering
DMA	differential mobility analyzer
DMB	1,2-diamino-4,5-methylenoxybenzene
DoE	design of experiments
DP	drug product
DS	drug substance
DSC	differential scanning calorimetry
DSMC	differential scanning micro-calorimetry
E. coli	*Escherichia coli*
EMA	European Medicines Authority
EndoH	endoglycosidase H
EndoS	endoglycosidase S
EPO	erythropoietin
ESI	electrospray ionization
Fab	antigen-binding fragment
Fabs	antigen-binding fragments
FDA	U.S. Food and Drug Administration
FFF	field flow fractionation

Continued on next page.

Table 1. (Continued). Acronyms

Acronym	Definition
FMEA	failure modes and effects
FTIR	Fourier transform infrared spectroscopy
Fv	fragment variable
GCSF	granulocyte colony stimulating factor
Gln	glutamine
GLP	good laboratory practice
GLP-Tox	Good Laboratory Practice-Toxicology
Gly	glycine
GMP	good manufacturing practice
GMP	good manufacturing practice
GPI	glycosylphosphatidylinositol
GS	glutamine synthetase
GS	glutathione synthase
H	heavy chain
H/D	hydrogen/deuterium
HCP	host cell protein
HDX	hydrogen-deuterium exchange
HIC	hydrophobic interaction chromatography
HILIC	hydrophilic interaction liquid chromatography
HMG	high-mannose glycan
HMW	high molecular weight
HPLC	high-performance liquid chromatography
ICH	International Conference on Harmonisation
icIEF	imaged capillary isoelectric focusing
IEF	isoelectric focusing
IEX	ion exchange chromatography
Ig	immunoglobulin
IM	ion mobility
INN	International Nonpropriertary Names
IPC	in-process control
ISO	International Organization for Standardization
ITAM	immunoreceptor tyrosine-based activation motif

Continued on next page.

Table 1. (Continued). Acronyms

Acronym	Definition
ITIM	immunoreceptor tyrosine-based inhibition motif
IV	intravenous
L	light chain
LALS	low-angle light scattering
LC-MS	liquid chromatography-mass spectrometry
LMW	low molecular weight
Lys	lysine
mAb	monoclonal antibody
mAbs	monoclonal antibodies
MALDI	matrix-assisted laser desorption/ionization
MALS	multi-angle light scattering
Man5	high-mannose glycan structure 5
Man9	high-mannose glycan structure 9
Met	methionine
MHC	major histocompatibility complex
MIRR	multi-chain immune recognition receptor
MoA	mechanism of action
mRNA	messenger RNA
MS	mass spectrometry
MSX	methionine sulfoximine
MTX	methotraxate
MW	molecular weight
NANA	N-acetylneuraminic acid
NANA	N-acetylneuraminic acid
Neu5Ac	*N*-acetylneuraminic acid
NGNA	N-glycolylneuraminic acid
NK	natural killer
NMR	nuclear magnetic resonance
nr	non-reduced
OPD	*o*-phenylenediamine
PAGE	polyacrylamide gel electrophoresis
PAT	process analytical technology

Continued on next page.

Table 1. (Continued). Acronyms

Acronym	Definition
PBMC	peripheral blood mononuclear cells
PCR	polymerase chain reaction
PD	pharmacodynamic
pI	isoelectric point
PK	pharmacokinetic
PNGase F	Peptide N-Glycosidase F
PPI	protein-protein interaction
PPQ	process performance qualification
PQA	product quality attribute
PQRA	Product Quality Risk Assessment
Pro	proline
PTM	post-translational modification
Q	quadrupole
QbD	quality by design
QC	quality control
r	reduced
R&D	research and development
RALS	right-angle light scattering
RF	rheumatoid factor
RM	NIST reference material
RP	reversed phase
SANS	small-angle neutron scattering
SAXS	small-angle X-ray scattering
scFv	single-chain fragment variable
SDS	sodium dodecylsulfate
SDS-PAGE	sodium dodecylsulfate-polyacrylamide gel electrophoresis
SEC	size exclusion chromatography
Ser	serine
SLS	static light scattering
SNP	single-nucleotide polymorphism
SpA	Staphylococcal protein A
SpG	Streptococcal protein G

Continued on next page.

Table 1. (Continued). Acronyms

Acronym	Definition
SRM	NIST standard reference material
SV	sedimentation velocity
SVA	sequence variant analysis
TEM	transmission electron microscopy
Thr	threonine
TNF	tumor necrosis factor
TOF	time-of-flight
tRNA	transfer RNA
Trp	tryptophan
TW	travelling wave
UHPLC	ultra-high pressure liquid chromatography
UPLC	ultra-high pressure liquid chromatography
USP	U.S. Pharmacopoeial Convention
V_H	variable domain of antibody heavy chain
V_L	variable domain of antibody light chain
WHO	World Health Organization

Table 2. Notes

Incorrect	Term	Correct
CE	cation exchange chromatography	CEX
CE-IEF	capillary isoelectric focusing	cIEF
CE-SDS	capillary sodium dodecylsulfate electrophoresis	cSDS
HX	hydrogen-deuterium exchange	HDX
SE-HPLC	size exclusion chromatography	SEC

Editors' Biographies

John E. Schiel

Dr. John E. Schiel received his B.S. (2004) and Ph.D. (2009) in Chemistry from the University of Nebraska-Lincoln, and is currently a research chemist in the NIST Biomolecular Measurement Division. He is leading the LC- and MS-based biomanufacturing research efforts at NIST; developing a suite of fundamental measurement science, standards, and reference data to enable more accurate and confident characterization of product quality attributes. Dr. Schiel is also the technical project coordinator for the recombinant IgG1κ NIST monoclonal antibody Reference Material (NISTmAb) program. He is an author of over 20 publications and recipient of numerous awards, including the ACS Division of Analytical Chemistry Fellowship, *Bioanalysis* Young Investigator Award, and UNL Early Achiever Award.

Darryl L. Davis

Dr. Darryl L. Davis holds a doctorate in Medicinal Chemistry from the Philadelphia College of Pharmacy and Science. His thesis focused on the use of MS in the characterization and quantitation of peptide phosphorylation. He started his career at J&J as a COSAT intern using MS to characterize the glycan linkages found on Remicade. Upon receiving his doctorate he accepted a full-time position within the Bioanalytical Characterization group at Centocor, a J&J company. Since joining J&J he has held a wide variety of responsibilities including starting and leading several sub-groups, analytical CMC lead, member of CDTs, member of technology development teams for alternative production platforms and new technology and innovation lead within analytical development. He has won several innovation awards within J&J for his work on automation and high-throughput analysis which continues to be a current focus. Currently he leads an analytical group within the discovery organization at Janssen R&D.

Oleg V. Borisov

Dr. Oleg V. Borisov earned a B.S. degree (with honors) in Chemistry at Moscow State University (1992), and received his Ph.D. in Chemistry from Wayne State University (1997), after which he completed his post-doctoral studies at Lawrence Berkeley National Laboratories (2000) and Pacific Northwest National Laboratories (2001). His background includes experience with analytical methods for characterization of biotherapeutic proteins and vaccine products, with emphasis on liquid chromatography and mass spectrometry methods.

© 2014 American Chemical Society

Dr. Borisov held positions at Genentech and Amgen with responsibilities that included protein characterization, testing improvement, leading innovation and CMC strategy teams. He is currently a Director at Novavax, Inc., developing methods and strategies for analysis and characterization of recombinant vaccines, based on nano- and virus-like particle technologies. His credits include several student awards, a book chapter, and over 25 scientific publications.

Indexes

Author Index

Barton, C., 69
Borisov, O., ix
Brorson, K., 99
Davis, D., ix, 1
Feng, J., 69
Flynn, G., 117
Harris, R., 69
Jefferis, R., 35

Kendrick, B., 99
Levitskaya, S., 69
Mire-Sluis, A., 1
Nyberg, G., 117
Schenerman, M., 69
Schiel, J., ix, 1
Spencer, D., 69

Subject Index

A

Acronyms, 151*t*
Allotype expression and amino acid correlates (Eu numbering)
  IgG1, 42
  IgG2, 43
  IgG3, 43
  IgG4, 43
  kappa and lambda light chains, 44

C

Capillary sodium dodecylsulfate electrophoresis (cSDS), 82
Chimeric and humanized antibody therapeutics, 5
cSDS. *See* Capillary sodium dodecylsulfate electrophoresis (cSDS)

G

Glycation, 88
Glycosylation, 88

H

Heterogeneity of IgGs. *See* Product attribute assessment
  analytical information, need, 70
  combined QC and extended characterization methods, 71
  critical quality attributes, identification, 72
  demonstrating product quality, 70
  product attribute assessment
  product attribute categories and tests, 74*t*
  qualitative profile assessment, 71
  technical challenges, 73
  testing plan based on product quality attributes, 72*f*
  use in diagnostics, reagents, and medicinals, 69

M

Monoclonal antibody therapeutics, 1
  antibody-drug conjugates (ADCs), 7
  chimeric antibodies, 5
  Fc fusion proteins, 7
  FDA-approved monoclonal antibody (mAb) and antigen-binding fragment (Fab) therapeutics, 8*t*
  humoral response, 3
  IgG-related therapeutic efficacy, 3
  introduction, 2
  mammalian cell culture, 6
  metrological reference material, 23
    complex drug products, evolution, 24
    national metrology institutes, 25
    NIST Biomanufacturing Program, 25
    USP medicines compendium, 24
  monoclonal antibody production
    downstream processing steps, 17*f*
    upstream processing steps, 15*f*
  monoclonal antibody (mAb) therapeutics, international nonproprietary naming, 4*t*
  murine epitopic determinants, 6
  NISTmAb IgG1κ, potential utility, 26
    biopharmaceutical design space, 27
    instrument or method performance, 27
    new techniques, suitability, 27
  production of mAb therapeutics, 14
    bulk drug substance, 14
    critical quality attributes (CQAs), 16
    drug product matrix, 16
    process analytical technology (PAT), 14, 18
    product-related impurities and substances, 14
    residual impurities, 15
    upstream and downstream processing, 14
  product-specific in-house reference standards, 19
    detailed characterization methods, 22
    method qualification and validation, 21
    primary and secondary reference, 20
    QC methods, 22
    representative monoclonal antibody lifecycle, 21*f*
  recombinant DNA technology, 5
Monoconal antibodies
  allotypes and idiotypes, 41

163

amino acid residues, non-covalent interactions with oligosaccharide, 49f
cellular IgG-Fc receptors
  FcγR binding sites on IgG, 53
  IgG-Fc receptors (FcγR) mediating antigen clearance, 51
chain structure for IgG1 molecule and inter-chain disulphide bridges, 39f
classical pathway, C1q/C1 binding and activation, 55
conclusions, 59
Fab, quaternary structure, 58
FcRn
  catabolism, 54
  transcytosis, 53
human antibody isotypes, 37
human antibody isotypes other than IgG, 59
human IgG
  allotypes, 42t
  gene polymorphism, 41
  polypeptide structure, 38
  quaternary structure, 45
IgG molecule, domain structure, 40f
IgG-Fc, quaternary structure, 45
IgG-Fc ligand binding, activation, and modulation, 51
IgG-Fc oligosaccharide
  moiety, 46
  sialylation, 57
IgG-Fc protein/oligosaccharide interactions, 48
influence of fucose and bisecting N-acetylglucosamine on IgG-Fc activities, 56
influence of galactosylation on IgG-Fc activities, 56
licensed chimeric mAb therapeutics, allotypy, 44
mechanisms of action, 35
recombinant IgG antibody therapeutics, IgG-Fc glycoform profiles, 50
representative IgG complex diantennnary oligosaccharides, 48f
role of IgG glycoforms in recognition by cellular FgRs, 55

P

Product attribute assessment
  charge
    deamidation and isomerization, 84
    sialylation, 85
  primary structure and post-translational modifications
    amino acid sequence fidelity, 87
    disulfide bonds, 87
    glycation, 88
    glycosylation, 88
    IgG1 and IgG4 disulfide bonding isoforms, 89f
    IgG2 disulfide bonding isoforms, 90f
    methionine and tryptophan oxidation, 92
  size
    aggregation, 79
    enzymatic fragmentation, 82
    fragmentation, 81
    non-enzymatic fragmentation, 82
    SDS-polyacrylamide gel electrophoresis (SDS-PAGE), 82
    total mass, 83
    truncation and extensions, 80
Product quality attributes, setting control strategy, 117
Protein moiety, 45

S

Sialic acid, 86
Sialylation, 86

U

U.S. Pharmacopoeial Convention (USP), 23
Using quality by design principles, 117
  advanced glycation end products (AGEs), 133f
  description of control elements, 119t
  detection (methodology), 125
  dimension, 126
  downstream detection, 126
  scoring matrix, 127t
  exposure to PQAs converting in Vivo, 147
  high-mannose attribute impact efficacy, 133
  mAbA high mannose, severity scoring, 134
  safety, 132
  impact of high-mannose glycans on drug clearance, 146
  likelihood of occurrence (process capability), 124

decision tree, 125f
mAbA glycation
  PQA assessment scoring, 142t
  PQRA scoring, 144t
mAbA high mannose, PQRA scoring, 127t
mAbB pharmacokinetic (PK), 148f
overall risk hazard, 129t
PQA changes in vivo study, 124f
PQA example 1, Fc HMGs
  detection, 136
  likelihood of occurrence, 135
  risk level determination, 137
PQA example 2, glycation
  detection, 143
  efficacy, 140
  likelihood of occurrence, 142
  mAbA glycation, severity score, 140
  risk level determination, 143
  safety, 139
PQA severity scoring
  efficacy subscores, 123t
  safety subscores, 122t
PQRA, lifecycle application, 130
preliminary risk hazard, 128t
product quality attribute (PQA) assessment, 121
product quality risk assessment (PQRA), introduction, 120
risk level, determination, 128
thioether concentrations, 148f
USP. See U.S. Pharmacopoeial Convention (USP)

W

Well-characterized biological proteins, perspectives
  analytical techniques
    extended physicochemical and biological characterization, 101
    release (and routine in-process testing), 101
    stability, 101
  biological characterization, 112
  biophysical characterization
    secondary structure, 109
    tertiary structure, 110
    thermal stability, 110
  charge variants, 106
  chemical modifications of select side chains, 100
  disulfide structure, 106
  elucidation of structure, overall conclusions, 112
  example characterization method summary, 103t
  forced degradation pathways, 110
  glycosylation, 105
  IgG2 and IgG4 complexity, 100
  new analytical methods, 102
  potential forced degradation conditions and degradation products, 111t
  PQA functional assessment, 114
  primary structure, 104
  product aggregates, 100
  product misfolding, 100
  reduction, 100
  relevant degradation pathways, recommended storage conditions
    chemical degradation, 113
    physical methods, 113
  size heterogeneity by cSDS, 108
  size heterogeneity by dSEC, 109
  size heterogeneity by SV-AUC, 108
  size variants and heterogeneity, size exclusion chromatography (SEC), 107
  structural and biochemical characterization, common strategies, 102